I0147940

THE LIVES OF
SNAKES

THE LIVES OF SNAKES

A NATURAL HISTORY OF THE WORLD'S SNAKES

Chris Mattison

PRINCETON UNIVERSITY PRESS
PRINCETON AND OXFORD

Published by Princeton University Press
41 William Street, Princeton, New Jersey 08540
99 Banbury Road, Oxford OX2 6JX
press.princeton.edu

Copyright © 2025 by UniPress Books Limited
www.unipressbooks.com

All rights reserved. No part of this book may be reproduced
or transmitted in any form or by any means, electronic or
mechanical, including photocopying, recording, or by any
information storage-and-retrieval system, without written
permission from the copyright holder. Requests for permission
to reproduce material from this work should be sent to
permissions@press.princeton.edu

Library of Congress Control Number 2024944437
ISBN 978-0-691-25060-1
Ebook ISBN 978-0-691-25061-8

Typeset in Bembo and Futura

Printed and bound in Malaysia
10 9 8 7 6 5 4 3 2 1

British Library Cataloging-in-Publication Data is available

This book was conceived, designed, and produced by
UniPress Books Limited
Publisher: Jason Hook
Project manager: David Price-Goodfellow
Designer & art direction: Wayne Blades
Picture researcher: Tom Broadbent
Illustrator: Robert Brandt
Copy Editor: Hugh Brazier
Maps: Les Hunt

Cover images: (Front): Kurit afshen/Shutterstock;
(spine and back): Chris Mattison

CONTENTS

What are snakes?

Snakes are creatures of curiosity, even amongst people who rarely if ever see them. Public opinion is probably more divided on the subject of snakes than any other group of animals. Though they are rarely seen, even in places where they are numerous, everyone seems to have an opinion on them, and it's usually a negative one. But before discussing the behavior of snakes, and the ways in which they interact with their environment, including ourselves, we first need to define what they are. Snakes are reptiles; in other words, they belong to the class Reptilia. The Reptilia consists of four orders: turtles and tortoises (Testudines), crocodilians (Crocodylia), tuataras (Rhynchocephalia), and the lizards, amphisbaenians, and snakes (collectively known as the Squamata).

Members of these orders are all tetrapod vertebrates (four-legged animals with backbones) covered with scales, which breathe through lungs and which derive their body heat from outside sources. Despite lacking legs, snakes are nevertheless tetrapods because they evolved from legged-lizard ancestors that had fully developed limbs. Their limbs were reduced and eventually disappeared through the course of evolution, although some snakes have pelvic girdles and vestigial limbs in the form of claw-like spurs on

either side of the vent (see pages 28 and 29). In some cases these are used by the males during courtship.

The order Squamata is further divided into three suborders: the lizards (Sauria), the amphisbaenians (Amphisbaenia; see page 16), and the snakes (Ophidia or Serpentes). Although it is convenient to refer to lizards and snakes separately, current information suggests the lizards do not necessarily have a common ancestor and that some families of lizards are more closely related to snakes than they are to other lizards. Snakes are therefore closely related to the lizards and amphisbaenians and less closely related to other reptiles. Separating snakes from legless lizards and amphisbaenians is not always straightforward. There are elongated, legless lizards in several families and, with three exceptions (the *Bipes* species from Mexico), amphisbaenians are also legless. In amphisbaenians, however, the squarish scales are arranged in rings.

EARLY SNAKES

The most widely held view is that snakes evolved from a group of lizards that took up a subterranean lifestyle, in which limbs became redundant or, indeed, a liability. Legless lizards are represented in at least seven families (see page 16).

The Anguidae family of lizards is placed in the modern suborder Anguinomorpha, which also includes the monitor lizards (Varanidae) and the beaded lizards (Helodermatidae). The prevalent view is that the snakes probably originated from a now-extinct group within the Anguinomorpha, even though they are now quite separate from the lizards. In addition, legless lizards have proportionately longer tails and shorter bodies when compared with snakes.

↑ Head of a typical colubrid snake, Green Trinket Snake, *Gonyosoma prasinum*, showing the lack of eyelids and ear openings, and the large plate-like head scales.

← Head of a typical lizard, a Dwarf Bearded Dragon, *Pogona henrylawsoni*, from Australia, showing the eyelids and the obvious ear openings, which are absent in snakes.

←← An amphisbaenian, Zarudny's Worm Lizard, *Diplometopon zarudnyi*, from the Middle East. The eyes of these poorly known reptiles are small and without eyelids, as in snakes, but the body scales are arranged in an annular pattern.

The earliest fossils that can be positively identified as snakes date from the Late Cretaceous period, about 70–95 million years ago. Snakes, however, are likely to have evolved earlier than this, probably during the Jurassic period, up to 150 million years ago. Snake fossils from these periods have been found but—frustratingly—they are incomplete, which is hardly surprising considering that snakes' skeletons and skulls consist of delicate, loosely articulating bones that easily become separated from each other after the animal dies.

AN IMPORTANT DISCOVERY

An important piece of the puzzle surfaced in 2006 when fossils that appear to represent a link between lizards and snakes, from the Late Cretaceous, were found in Argentina. This new species was given the name *Najash rionegrina*, and it had a well-developed pelvic girdle and the remains of well-formed, robust and functional hind limbs, like some forms of living lizards. At the time, the skull was too deformed to provide much information, but in 2013 additional fossils were found and these included a complete skull showing similarities to the skulls of modern snakes,

identifying *Najash* as a link between lizards and early snakes. Despite this find, much remains to be investigated before the ancestry and lifestyle of these reptiles can be fully uncovered.

THE LARGEST KNOWN SNAKE

The extinct species *Titanoboa cerrejonensis*, described in 2009, is the largest snake known. It has no widely accepted common name, but the genus name translates literally as "Titanic Boa."

Titanoboa lived 58–60 million years ago, during the Paleocene epoch. Its fossils were discovered in present-day northern Colombia, from the huge open-cast Cerrejón coal mine, where the remains of about 30 individuals have been found. These consist mostly of individual vertebrae and ribs, from which the overall size of the snake has been extrapolated to approximately 43 ft (13 m) in length and 1.25 tons in weight. By comparison, the largest living snake, the Green Anaconda, *Eunectes murinus*, averages about 22 ft (6.6 m) in length—roughly half of that estimated for *Titanoboa*.

← One of many legless lizards, FitzSimons' Burrowing Skink, *Typhlacontias brevipes*, from the Namib Desert. Members of the skink family (Scincidae) show a tendency toward limb reduction or loss.

←← A Blue-tailed Monitor, *Varanus doreanus*. Snakes are thought to have arisen from the same evolutionary branch as the monitor lizards (Varanidae), the family that also includes the huge Komodo Dragon, the world's largest lizard.

→ The Green Anaconda, *Eunectes murinus*, the world's heaviest snake and thought to be a distant relative of even larger, extinct species.

↓ A Green Anaconda, *Eunectes murinus*. It is semi-aquatic, which helps support its bulky body.

Its relationships with modern snake families place it in the Boidae, and the subfamily Boinae. Its closest living relatives, therefore, are the large boas, which include the anacondas and the Common Boa, *Boa constrictor*.

From its great size, and some of the fossils of other organisms with which it was found, it is likely that *Titanoboa* lived in shallow water and muddy swamps. In the Paleocene period the Cerrejón region was located within a large tropical rainforest inundated with huge river systems, deltas, and shallow lakes. *Titanoboa*'s large body size would have made it difficult or impossible for adults to live and move about on the land, whereas an aquatic or semi-aquatic habitat would have helped to support its huge body. The modern-day anacondas have adopted a semi-aquatic lifestyle for the same reason. Whereas anacondas, however, have a diet consisting largely of crocodilians and mammals, *Titanoboa* may have eaten fish. A single specimen has been found in which the skull is present, and it is similar to the skulls of modern fish-eating snakes. If this is the case, *Titanoboa* would be unique amongst the Boidae in being piscivorous.

← *Titanoboa cerrejonensis*, a snake that lived 58–60 million years ago, weighed up to 1.25 tons and grew as long as 43 ft (13 m).

How are snakes classified?

Modern snake species are divided into two groups, or clades: the Scolecophidia (five families of worm snakes) and the Alethinophidia (true snakes). The Alethinophidia is further subdivided into the Amerophidia, whose members are restricted to Latin America, and the Afrophidia, which evolved in Africa and subsequently radiated out into other parts of the world.

↙ The South American Pipe Snake, *Anilius scytale* (top), is the sole member of its family, the Aniliidae, belonging to the Amerophidia. The Spotted House Snake, *Alopecion guttatum* (bottom), a rarely seen species from Namaqualand, South Africa, is a member of the afrophidian family Lamprophiidae, which is restricted to Africa and the Seychelles.

The Afrophidia is subdivided in turn, into two groups, the Henophidia and the Caenophidia, the "old" snakes and the "recent" snakes, respectively. The Henophidia contains the boas, pythons, pipe snakes, shield-tails, and two small families of sunbeam snakes. The Caenophidia contains the file snakes—three species belonging to the extant family Acrochordidae and two other families known only from fossils—and all the remaining snakes, classed together as the Colubroidea and consisting of eleven families and over 3,000 species, making it by far the most numerous group. Included within it are the ratsnakes, whipsnakes, tree snakes, water snakes, cobras, vipers, sea snakes, and many others.

The arrangement of snakes into families and superfamilies has changed many times over the last couple of decades and will probably continue to do so as more information becomes available. At present most experts accept the basic division into two "infraorders," the Scolecophidia (worm snakes) and the Alethinophidia (all the rest), as explained above. Some debate concerns the latter group, however, which is further subdivided but not in a way that is unanimously agreed. The table opposite is therefore a snapshot of the current situation, which is fluid, and which may change again before settling down into a generally accepted conclusion, especially where the relationships between the very earliest snakes, the Scolecophidia, are concerned.

CLASSIFICATION OF MODERN SNAKES, SHOWING NUMBER OF SPECIES IN EACH FAMILY

INFRAORDER	CLADES	SUPERFAMILY	FAMILY	NO. OF SPECIES IN FAMILY
Scolecophidia			Typhlopidae	281
			Leptotyphlopidae	143
			Anomalepidae	20
			Gerrhopilidae	29
			Xenotyphlopidae	1
Alethinophidia	Amerophidia		Aniliidae	1
			Tropidophiidae	38
	Afrophidia Henophidia		Xenophidiidae	2
			Bolyeriidae	2
			Anomochiliidae	3
			Cylindrophiidae	15
			Uropeltidae	64
			Loxocemidae	1
			Pythonidae	38
			Xenopeltidae	3
			Boidae *	67
	Caenophidia	Acrochordoidea	Acrochordidae	3
		Colubroidea	Xenodermidae	36
			Pareatidae	44
			Viperidae	383
			Homalopsidae	57
			Cyclocoridae	8
			Pseudaspididae	4
			Elapidae	400
			Prosymnidae	18
			Psammophiidae	57
			Lamprophiidae	90
			Pseudoxyrhophiidae	88
			Atractaspididae	69
			Colubridae **	2,119
Total				**4,084**

* The Boidae is sometimes split into seven separate families—Boidae, Calabariidae, Candoidae, Charinidae, Erycidae, Sanziniidae, Ungalioiphiidae—but here they are retained as subfamilies.

** The Colubridae is also sometimes split into seven separate families—Calamariidae, Colubridae, Dipsadidae, Grayiidae, Natricidae, Pseudoxenodontidae, Sibyophidae—but here they are retained as subfamilies.

Where do the scientific names come from?

All living organisms have a two- or three-part scientific name consisting of the genus, which always begins with an uppercase letter, and the species, which begins with a lowercase letter (even if it is a proper noun, such as somebody's name). Sometimes, a subspecific name, which also begins with a lowercase letter and may be a repeat of the specific name, is also used—although the subspecies concept is not accepted by all experts.

← Schultze's Pit Viper, *Trimeresurus schultzei*, a spectacular viper from the Philippines, is named after Dr. Leonhard Schultze-Jena (1872–1955), a German geographer, zoologist, and botanist.

↗ One of just seven snakes with tautonyms, this is a juvenile Mussurana, *Clelia clelia*, a South American species which loses its spectacular coloration as it matures.

→ The Crowned Dwarf Snake, *Eirenis coronella*, has a generic name that comes from the Greek goddess of peace, Eireni (Irene), presumably referring to the docile nature of this species, while *coronella* means small crown.

TAUTONYMS:
SO GOOD THEY NAMED IT TWICE

Tautonyms are names in which the generic name is
repeated as the specific name. There are currently seven
examples among snakes: *Cerastes cerastes*, the Desert
Horned Viper; *Clelia clelia*, the Mussurana; *Enhydris
enhydris*, the Rainbow Water Snake; *Hypnale hypnale*,
the Hump-nosed Viper; *Natrix natrix*, the Common Grass
Snake; *Naja naja*, the Common or Spectacled Cobra;
and a sea snake, *Suta suta*, sometimes known as the
Curl Snake. A few species have names that are almost
tautonyms, such as the Rinkhals or Ring-necked Spitting
Cobra, *Hemachatus haemachatus*.

How do scientists arrive at the names they give to
new species? There is no set rule, and names may be
descriptive of the snake, its habits, habitats, or the place
where it was first discovered, or it may honor the
person who discovered it or someone who advanced
the science of herpetology.

Other names are derived from classical characters,
and some are assigned arbitrarily, at the whim of the
person who first described the species. Looking at
a few examples, many snakes have the specific name
lineatus, meaning "striped" or "lined," whereas *fasciatus*
means "banded." The generic name of the larger
rattlesnakes, *Crotalus*, is derived from the Greek *krotalon*,
meaning a little bell or rattle. Geographical names
include *Eirenis africanus*, a small colubrid snake from
Africa, and *Leptophis mexicanus*, the Mexican Parrot
Snake, while four species have the specific name
capensis, for the Cape region of South Africa, and five
are named *madagascariensis*.

Finally, classically derived names include *Python*,
the mythical serpent slain by Apollo, and *Naja*, from
Hindu mythology. Certain venomous snakes reflect the
names of the Three Fates of Greek mythology: Clotho,
Lachesis, and Atropos. Clotho spins the thread of life
and was an early name, now obsolete, for some of the
vipers belonging to the genus *Bitis*; Lachesis measures
the length of the thread and gives her name to the
genus of bushmasters; while Atropos—whose name
occurs in the Berg Adder, *Bitis atropos*—was the Fate
who cut the thread.

Snake look–alikes

In an example of convergent evolution, some lizards have lost their limbs through the process of evolution. Without exception, these are burrowing or semi-burrowing species, and they belong to a number of different families, showing that leglessness has arisen more than once.

Legs clearly get in the way when an animal is burrowing through the ground and, as it has no use for them, there will be selective pressures for them to become reduced in size, sometimes with fewer bones, and eventually to disappear altogether, at least outwardly.

Amphisbaenians are a poorly known group of legless reptiles that are restricted to warmer regions and spend their whole lives underground. They resemble snakes, but only superficially; the arrangement of their scales in rings is very distinctive. Legless lizards are found in the Scincidae (skinks), Anguidae (glass lizards and slowworms), Dibamidae (blind skinks), and Anniellidae (American legless lizards). Three other families have at least some species with greatly reduced limbs: Gymnophthalmidae (spectacled lizards or microteiids), Cordylidae (girdled lizards), and Pygopodidae (flap–footed lizards). Between them, these seven families are found almost worldwide.

→ Five reptiles that resemble snakes:
(A) Sandfish, *Scincus scincus*, a sand-swimming skink that folds its limbs against its body when it is moving through sand. (B) Striped legless skink, *Acontias lineatus*. (C) An Australian Burton's Flap-footed Lizard, *Lialis burtonis*, one of the Pygopodidae, whose members all retain only rudimentary limbs. (D) Slow-worm, *Anguis fragilis*, with young: a legless European member of the Anguidae. (E) Spanish Worm-lizard, *Blanus cinereus*, an amphisbaenian.

Diversity and adaptation

At the latest count there were just over 4,000 species of snakes. Despite a similar body plan, consisting of a more-or-less cylindrical shape without limbs, eyelids, or external ear openings, there is enough variation between them to separate all of these species.

Differences between some of them are slight but, by and large, most interested people would be able to identify the snakes they are likely to come across in their own locality—as well as many that occur further afield. More importantly, the snakes themselves are able to recognize members of their own species, thereby limiting the risk of accidental hybridization.

Species differ not only in their size, shape, color, and markings, but also in the arrangement of their scales (their scalation). In addition, snakes use chemical communication, in the form of smell, to recognize members of the same species, and this is their most important means of species and gender recognition. These variations and combinations of characteristics have not occurred by accident but are the results of millions of years of evolution, the fine-tuning that takes place to make each generation more "fit" to survive under the conditions in which they live.

EVOLUTIONARY PRESSURES

Over the millennia, snakes found themselves in new places where they were subject to a variety of evolutionary pressures: to become bigger, smaller, faster, slower, darker, lighter, and so on, to stand a better chance of surviving and passing on their genes. This, together with changing conditions over long periods of time, and competition with other members of the same or other species, was the driving force behind numerous adaptations and fine-tuning, resulting in the ensemble of species present today in any given place. Competition between species can be avoided, or minimized, by occupying different niches—evident in different prey types, activity

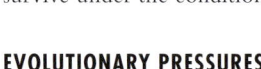

patterns, habitat preferences, and so on. This is especially important in parts of the world with large numbers of species, such as tropical rainforests, where competition is fierce.

← Blackneck Garter Snake, *Thamnophis cyrtopsis*, one of the more colorful members of its genus, which also includes the ribbon snakes.

→ A tree snake, *Imantodes lentiferus*, from South America, a remarkably slender snake popularly known as the "bootlace snake."

↓ The Mandarin Rat Snake, *Euprepiophis mandarinus*, among the most extravagantly patterned colubrid snakes.

Global distribution

Snakes are not evenly distributed throughout the world but are more numerous in some places than others. Several factors are involved—not only the availability and suitability of habitats, but also, importantly, the physical ability of snakes to spread overland, which is likely to be slowly.

→ The Pacific Ground Boa, *Candoia carinata*, is from the Pacific region. This species, and four others in the same genus, are separated from other boas by many thousands of miles and are placed in a separate subfamily, the Candoiinae.

↓ *Trachyboa boulengeri*, a secretive and poorly known member of the Tropidophiidae from South America.

Unlike many birds, snakes are unable to cross large expanses of sea, and most are equally unable to cross extreme deserts or mountain ranges. These barriers may not appear until after a certain amount of dispersal has already taken place, with the result that populations that made those journeys beforehand become isolated. In this way, new variations and, ultimately, new species appear.

Distribution on a global scale provides clues about the relationships between snake families and the timescales of their appearance: some families evolved long ago and spread out when the major landmasses were joined, allowing new areas to be colonized relatively easily. Families that were initially widespread, however, were sometimes superseded in places by members of more successful families. Others evolved

later, after some landmasses had drifted apart, restricting their dispersal and resulting in a more limited distribution. Snakes that become isolated on smaller landmasses, such as oceanic islands, have very few opportunities to spread or to be "refreshed" by new individuals and may, in time, evolve into new species not found on the nearby mainland.

Superimposed on this pattern is one of habitat preference. Different snakes tend to live in different habitats, and so the present-day distribution of species and families is an interplay between the practicalities of dispersal, coupled with the availability of suitable habitats. As a rule, generalist species will spread and colonize new areas more easily than specialists.

Size and shape

Snakes are unable to generate their own body heat internally and must rely on outside sources—which ultimately means the heat of the sun. This has a bearing on their size and shape: large bodies take longer to warm up than small ones, so the largest snakes live in parts of the world where ambient temperatures are close to their preferred level, which is about 85 °F (30 °C).

Large snakes are also likely to occupy open habitats such as grasslands, swamps, or riverbanks, rather than closed-canopy forests. Burrowing and aquatic snakes are similarly restricted to warm regions, because they have limited opportunities to raise their body temperatures above that of their surroundings by basking or by shuttling between cool and warm places.

In addition, body shape depends to a large extent on lifestyle, so arboreal species are mostly long and thin so that boughs can support their weight,

← The Reticulated Python, *Malayopython reticulatus*, from the rainforests of Southeast Asia, is generally considered to be the longest species of living snake, although the Anaconda, *Eunectes murinus*, being stouter, is the heaviest.

↑↑ A small worm snake from Sri Lanka, possibly an undescribed species belonging to the family Typhlopidae. The small fossorial snakes belonging to this and closely related families contain the smallest species.

↑ Namaqua Dwarf Adder, or Schneider's Adder, *Bitis schneideri*, the world's smallest viper and one of the stoutest snakes in the world. It has a limited range in southern Africa.

whereas burrowing species are more likely to be short and stout so that they can drive themselves through the substrate. Snakes that ambush their prey are more heavy-bodied than the species that chase down their prey.

THE LARGEST

The two largest snakes are the Green Anaconda, *Eunectes murinus*, from South America and the Reticulated Python, *Malayopython reticulatus*, from Southeast Asia. The largest recorded Anaconda, shot in 1907, was 62 ft (18.9 m) long, although there is considerable doubt over the accuracy of this record. A more realistic maximum size would be around 33 ft (10 m), and there are a number of records of Anacondas reaching this length.

There are several records of Reticulated Pythons growing to almost 30 ft (9 m) in captivity, and wild individuals may approach this length. Other "giant" species include the Burmese Python, *Python bivittatus*, and the closely related Indian Python, *P. molurus*, which can each potentially grow to about 20 ft (6 m), and the African Python, *P. sebae*, of similar size or slightly smaller. The Common Boa, *Boa constrictor*, is, perhaps disappointingly, much smaller than any of these, at a maximum of just over 13 ft (4 m). Having said that, coming across even a 13 ft snake unexpectedly makes the heart beat faster—and meeting one 20 ft long would be an awesome encounter.

THE SMALLEST

The smallest snake in the world is probably the Barbados Thread Snake, *Tetracheilostoma carlae*, described (as *Leptotyphlops carlae*) in 2008. Adults measure about 4 in (10 cm) in length. Several other thread snakes are only slightly larger, and there may be smaller species waiting to be discovered. A snake of 4 in is tiny, and its body has to contain all the organs, muscles, and skeletal parts that it needs to live. Its eggs are smaller than a grain of rice, and hatchlings are truly "thread-like." I can think of no other group of animals in which the difference in size, and mass, between the largest and the smallest is so great.

FORM AND
FUNCTION

Skeleton and skull

Compared with their nearest relatives, the lizards, snakes have a greatly modified skeleton, consisting simply of a skull and a number of vertebrae and ribs. The vertebrae are divided into two types: the precaudal (in front of the tail) and caudal (tail). In total there may be up to 400 vertebrae, depending on species.

Each precaudal vertebra has a pair of ribs attached, but there is no breastbone so the free ends of the ribs are connected to each other by muscles. These are used to control the vertebrae during locomotion and, in some species, constriction. The caudal vertebrae lack ribs and are graduated in size, reflecting the tapered shape of the tail.

Each vertebra has a cylinder-shaped central portion with one convex and one concave end so that they fit together and articulate with each other. There is an arch of bone on top of each vertebra (the neural arch) through which the spinal cord passes.

→ Skeleton of a coiled python, showing the numerous pairs of ribs and their associated vertebrae, which can number over 400 in some species.

Snake skull
The skull of a typical snake, showing the teeth and the extremely loosely attached bones.

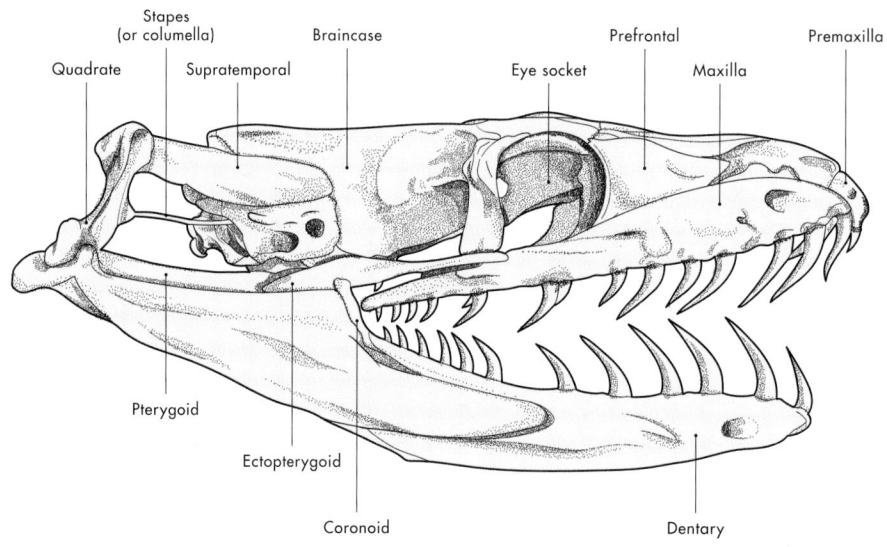

Quadrate · Stapes (or columella) · Supratemporal · Braincase · Eye socket · Prefrontal · Maxilla · Premaxilla

Pterygoid · Ectopterygoid · Coronoid · Dentary

There are also short spines attached to the cylinder of each vertebra, and these interlock loosely with those from neighboring vertebrae; the purpose of these is to limit the amount by which each vertebra can "spin" relative to its neighbors, and thus prevent damage to the spinal cord (see page 180).

Downwardly pointing processes are also found on the bottom of each vertebra in most advanced snakes—though not in the oldest families, nor in some burrowing snakes. Their function is not altogether clear, although in the egg-eating snakes, *Dasypeltis* and *Elachistodon*, these processes are greatly enlarged in the

esophageal region, extending into the top of the throat where they are used to saw through the shells of the birds' eggs on which they feed.

No snakes have pectoral girdles, the part of the skeleton to which the front limbs are attached in limbed animals, but members of the oldest families, such as the blind snakes and thread snakes, have the

Schinz's Beaked Blind Snake, *Rhinotyphlops schinzi*, from South Africa, is a member of one of the most primitive families, the Typhlopidae.

Snakes belonging to the Boidae, and some other primitive families, have small spur-like structures attached to vestigial pelvic girdles.

A Cuban Wood Snake, *Tropidophis melanurus*, a member of the Tropidophiidae, restricted, like several others in this small family, to Cuba.

remnants of pelvic girdles, as do members of several other families of primitive snakes, collectively known as henophidians. These include pipe snakes, shield-tailed snakes, pythons, and boas. In the boas and pythons, Boidae and Pythonidae, and perhaps in some of the other, lesser-known families, small spur-like processes are associated with the pelvic girdle, and these may be used to stimulate females during courtship, apparently their only known function. In keeping with this, the spurs of males are often proportionately larger than those of females.

Teeth and fangs

For animals that have no limbs, teeth and fangs are the "tools of the trade" when it comes to subduing prey, and they are also used as a means of defense. The arrangement of teeth and fangs reflects the lifestyle of each species, as well as its underlying relationships. A great deal can be deduced about the way in which a snake feeds—and its preferred prey—from a study of its dentition.

There is much variation in the number and arrangement of teeth between species. Some snakes, belonging to the Scolecophia, have almost no teeth at all, whereas others have numerous teeth—which may be of several different types. The teeth are situated along the edge of the lower jawbones (the maxilla), the pterygoid bones, and the palatine bones. They are attached to the inside edges of the bones (pleurodont) as opposed to being fixed in sockets (thecodont) as they are in some other reptiles, and in mammals.

Teeth are replaced throughout the snake's life by new ones that develop at the bases of the old ones, ready to move into place whenever an old tooth is lost. Shed teeth, including venom-injecting fangs, are often swallowed along with the prey in which they are embedded, and they can sometimes be found in snakes' feces.

← The jaws of typical snakes, such as this Central American Parrot Snake, *Leptophis praestans*, contain several rows of teeth of differing shapes and sizes, depending on the relationships and feeding habits of the species concerned.

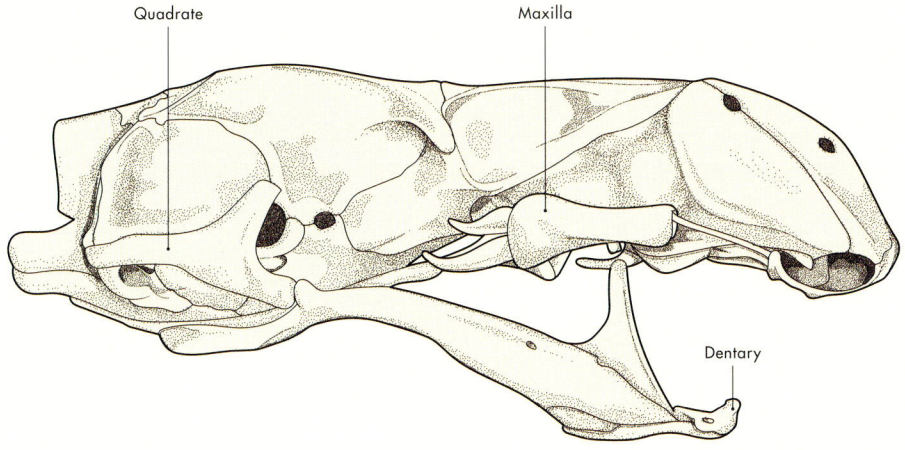

Quadrate   Maxilla

Dentary

REAR FANGS

Typical snakes, such as most colubrids, have teeth that are almost uniform in shape and size. Other species have enlarged teeth at the rear of their mouths, and these are commonly known as "back-fanged" or "rear-fanged" snakes. The fangs are associated with Duvernoy's glands, which are modified salivary glands. A duct carries toxic saliva—in effect venom—to the bases of the enlarged rear fangs, usually a single pair but sometimes two or three pairs, depending on species. These rear-fanged snakes need to hold and chew their prey in order to allow the venom to run along a groove in their fangs and enter a wound. With a few notable exceptions, rear-fanged snakes are not considered dangerous to humans.

Absence of fangs in lower jaw

The skull of a member of the Typhlopidae. These snakes have no teeth in their lower jaws and one or two in the upper jaw, but their skulls are more flexible than those of the Leptotyphlopidae (see page 155).

Enlarged rear fangs

Enlarged rear fangs are present in members of several families. Venom may enter the wounds made by these cutting fangs, and some species are dangerous to humans.

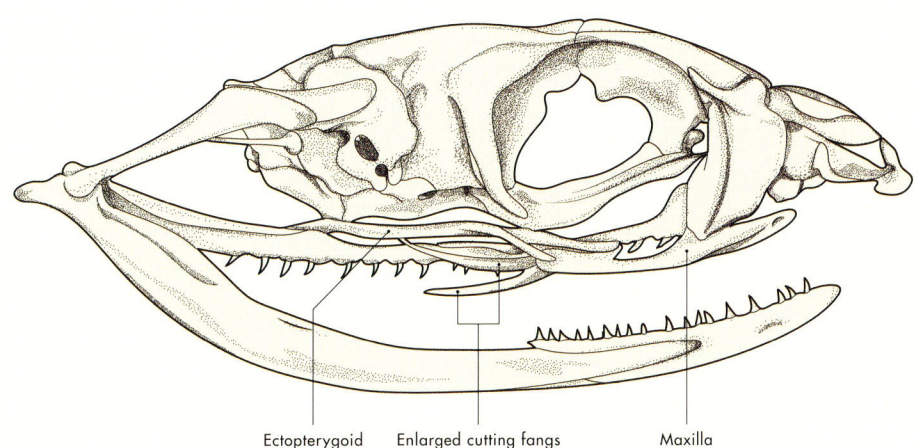

Ectopterygoid   Enlarged cutting fangs   Maxilla

VENOMOUS FRONT FANGS

Front-fanged snakes, members of the Elapidae and the Viperidae, have specialized teeth at the front of their mouths. Elapids (cobras, mambas, kraits, coral snakes, and their relatives) have relatively short, fixed fangs, whereas those of the vipers are hinged, so that they can be folded against the roof of the mouth when not in use; this allows them to be longer and therefore to penetrate more deeply into the prey's body.

The fangs of the Gaboon Viper, *Bitis gabonica*, and the West African Gaboon Viper, *B. rhinoceros*, measure up to 2 in (5 cm) in length, and are usually considered the longest fangs of all. Large individuals of Eastern and Western Diamondback Rattlesnakes, *Crotalus adamanteus* and *C. atrox*, have fangs that are only slightly smaller.

Elapid and viper fangs are hollow, allowing the venom to travel to their tips before being injected into the prey. The venom is pumped to the fangs by contractions of the masseter muscles surrounding the venom sac. This forces it along venom ducts leading to the base of the fangs and into the hollow fang, where it emerges via a small aperture near the tip.

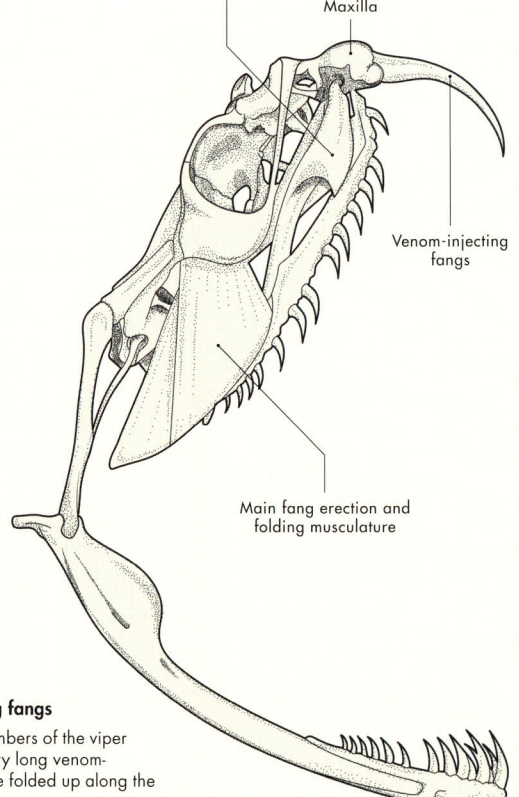

Ectopterygoid

Maxilla

Venom-injecting fangs

Main fang erection and folding musculature

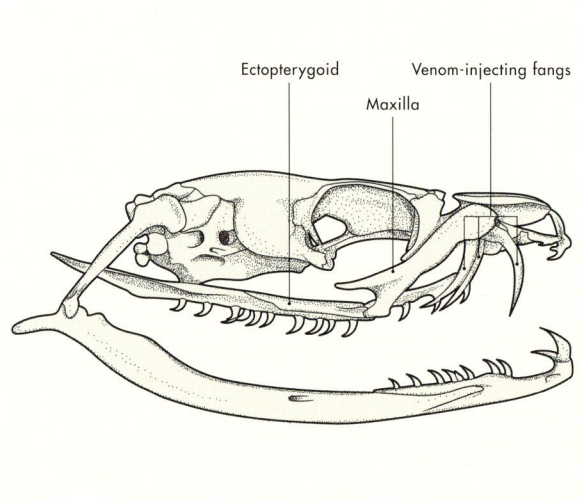

Ectopterygoid

Maxilla

Venom-injecting fangs

Non-erectile, forward-mounted fangs

Fixed and enlarged hollow front fangs for injecting venom, as in cobras, coral snakes, mambas and other elapids.

Erect venom-injecting fangs

The upper jaws of members of the viper family, Viperidae, carry long venom-injecting fangs that are folded up along the upper jaw when not in use.

VARIATIONS ON A THEME

There are some variations from these basic plans. Snakes belonging to the genus *Atractaspis* are known as stiletto snakes due to their long, hollow fangs. Like vipers' fangs, those of stiletto snakes can be folded up against the roof of the mouth but, unlike vipers, they can be swung sideways to emerge from the side of the mouth, so the snake can bite (or, more accurately, "stab") while its mouth is partially closed. This enables it to use them in narrow underground spaces, hunting the small burrowing mammals and reptiles on which it feeds.

Specialized teeth are found in other groups of snakes, for instance the enlarged and serrated teeth that occur in the so-called kukri snakes, *Oligodon* species, from Asia: these probably enable the snakes to slice through the tough leathery shells of the reptile eggs on which they often feed. Other species, notably fish-eating snakes such as the file or wart snakes, *Acrochordus*, and the water snakes belonging to the Homalopsidae, have teeth with striations or flutings, which help them to penetrate the scaly skins of fish.

Backward-facing fangs

Some members of the Atractaspididae have long, hollow, venom-injecting fangs that can be deployed without fully erecting them, giving them the popular name of stiletto snakes.

↖ The long fangs of a rattlesnake, *Crotalus* species, that have been lowered with a probe to show their shape and length. When they are folded up against the roof of the mouth they are covered with a fleshy sheath, which can be seen here, partially folded back.

↑ Bibron's Stiletto Snake, *Atractaspis bibroni*. Members of this genus are popularly known as stiletto snakes. They hunt underground, where opening their jaws widely to inject venom would be impractical.

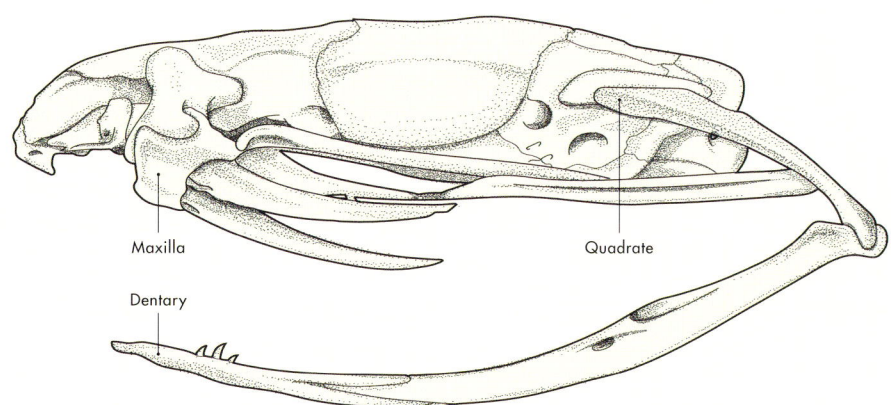

Maxilla

Quadrate

Dentary

Skin and scales

The skin between the scales, known as the interstitial skin, is normally a uniform gray. It is usually hidden unless the snake's body is distended when swallowing large prey or when it contains a large clutch of eggs or developing young. Some species expand the neck region when displaying aggression, and this may expose underlying bold patterns, as in the Southern Twig Snake, *Thelotornis capensis*.

Snake head scales nomenclature

The heads of many snakes from several families are covered with large plate-like scales, which can often be useful in identifying them.

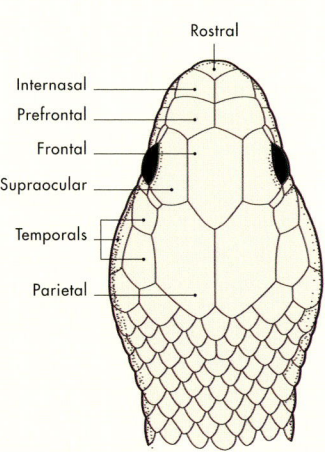

DORSAL VIEW

Rostral

Internasal

Prefrontal

Frontal

Supraocular

Temporals

Parietal

LATERAL VIEW

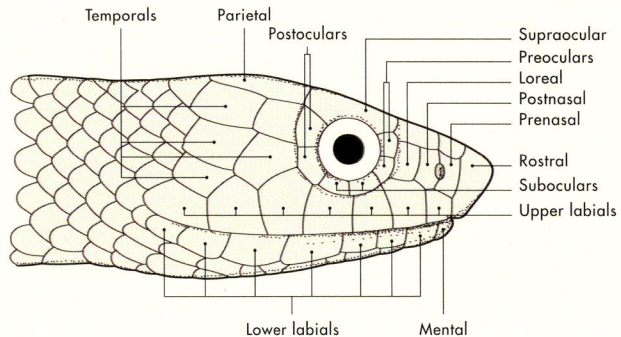

Temporals

Parietal

Postoculars

Supraocular

Preoculars

Loreal

Postnasal

Prenasal

Rostral

Suboculars

Upper labials

Lower labials

Mental

THE STRUCTURE OF SCALES

Scales are formed from thickened areas of skin. Different types of scales are found on different parts of the snake (except in the most primitive species). They may be roughly divided into head scales, which may be large and plate-like or small and granular, dorsal body scales, of which there are several rows and which overlap like the tiles of a roof, and ventral scales, which are arranged as a single overlapping wide row. The ventral scales under the tail (subcaudal scales) may be in a single or double row. The types of scales often reflect the snake's lifestyle: burrowing snakes usually have smooth scales to help them slip through the soil or sand, whereas aquatic snakes often have heavily keeled scales to give them greater water resistance, helping them to push forward.

→ This Large-eyed Green Treesnake, *Rhamnophis aethiopissa*, a spectacular rear-fanged arboreal snake from Africa, has puffed up its throat to display the bold blue and green markings, outlined by black interstitial skin, intended to intimidate.

→→ The scales on the flanks of a Sidewinder, *Crotalus cerastes*, are roughened by the presence of a raised keel down the center of each scale (top inset). Baird's Ratsnake, *Pantherophis bairdi*, is covered in smooth scales. The paired apical pits (see page 50) can clearly be seen in this close-up (bottom inset).

Snake undertail scales nomenclature

The scales underneath snakes' tails are known as subcaudal scales. They may be single (upper figure) or paired (lower figure), depending on the species. Tails of males are usually longer, and have a broader base, than those of females.

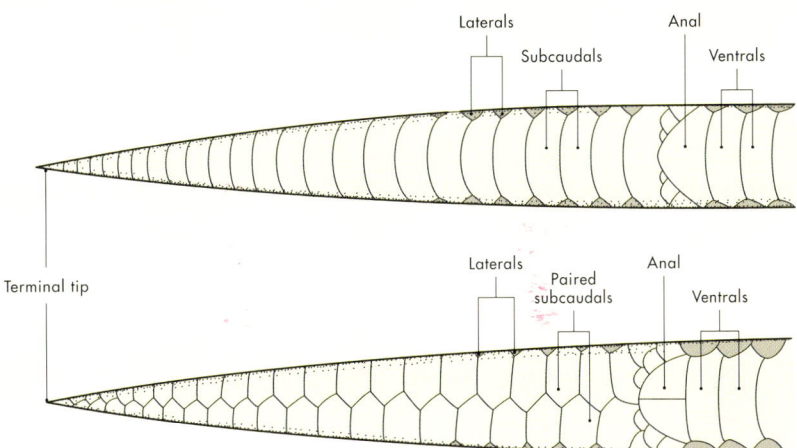

Dorsal scales vary in size, shape, color, and texture, depending on species. In a few species, such as the Hairy Bush Viper, *Atheris hispida*, each keeled scale may be elongated into a point producing a prickly or spiny effect. A small number of species, notably the three file snakes of the family Acrochordidae, and the strange Dragon or Javan Tubercle Snake, *Xenodermus javanicus*, have rough, granular scales, thought to improve their grip when they are subduing the slippery fish and eels (and frogs in the case of the latter species) on which they feed.

← The scales of the Hairy Bush Viper, *Atheris hispida*, are drawn out into points, giving the impression that the snake is covered in spines.

→ This Adder, *Vipera berus*, has secreted a layer of oil between the old and new epidermis to loosen the upper layer so that it can be sloughed off.

↓ A section of a Catesby's Snail-eater, *Dipsas catesbyi*, that is shedding its skin.

SHEDDING

Whereas color cells, or chromatophores, are contained in the underlying part of the skin, known as the dermis, the upper layer, the epidermis, is more-or-less transparent—and it is this part that is shed occasionally. Shedding occurs more often in young, rapidly growing snakes than in adults, which may only shed two or three times each year.

In temperate species that hibernate, shedding usually occurs shortly after they have emerged in the spring, and this is the time females are most attractive to males. Female snakes also shed prior to egg-laying or birth, and the timing of this is often predictable: for example, the pre-laying shed takes place about 8–10 days before laying in the case of milksnakes, kingsnakes, and ratsnakes, but other species may be different.

Shedding begins with the secretion of an oily substance between the old and new layers of epidermis, resulting in the snake's eyes becoming cloudy or "blue." A few days later the eyes clear and the snake initiates shedding by rubbing its snout on a rough surface until the skin is broken. It then begins to crawl out of its old skin, which is turned inside-out in the process. Shed skins are often found entangled in vegetation or rocks.

THE RATTLESNAKE'S RATTLE

Rattlesnakes' rattles are formed as a result of part of the shed skin not being completely discarded. This occurs because the end scale, instead of being conical as in most snakes, has a small bulbous swelling at its tip, known as the "button." Newborn rattlesnakes have no segments, just the button, but each time they shed, the section over the button comes loose but cannot be completely freed because of its shape, and so it remains loosely attached, surrounding the button, while the rest of the shed skin is torn away. Each subsequent shed adds another segment until there is a rattle, slightly tapered in shape because each shed skin is larger than the one before it.

When aggravated, a rattlesnake vibrates its tail, thereby causing the rattle to shake. In time, the rattle becomes brittle and the furthest (oldest) section breaks off through wear and tear, leaving fewer segments and a more parallel-sided rattle. Most rattles on wild snakes consist of about six or seven segments, but captive rattlesnakes, which lead a relatively passive life, may accumulate 10 or more segments.

→ A Sidewinder, *Crotalus cerastes*, showing its rattle. Note also the tongue "tasting" the air and the beautifully camouflaged coloration.

↓ Two rattlesnake rattles. The upper one is from an old snake, and the segments are of a similar size, with earlier segments broken off. The lower one is from a young snake that is still growing, with the segments becoming progressively larger.

Snakes of many colors

Snakes' colors are largely dependent on the environment in which they live. In cool places, dark-colored snakes will warm up more quickly when exposed to radiant heat from the sun, whereas light-colored species reflect heat. The need to maintain the right temperature must be balanced with the need to be inconspicuous and avoid the attentions of predators.

DIFFERENT TYPES OF COLOR

Colors in snakes are produced in three different ways. Pigments may be contained within the color cells (chromatophores). These consist of colored chemicals, of which melanin is the most common, resulting in brown or black areas. Red and orange are produced by carotenoids, and yellow may be produced by either melanin or carotenoids; white is produced by guanine.

Iridescent effects are the result of a process known as interference. This occurs when light strikes the translucent outer layer of a snake's scales at an angle and splits into its component parts, like a rainbow or a film of oil on a puddle.

Finally, there is Tyndall scattering, in which small particles known as iridophores embedded in the snake's cells refract and reflect light in a particular way, so that light at the shorter (blue) end of the spectrum is reflected more than other parts. Very few snakes are blue, however, because in most cases a layer of yellow chromatophores overlies the blue-producing cells, turning the blue to green—although blue snakes do exist.

SHADES AND PATTERNS

More than one type of color production may be found in a single snake, thus forming various shades and patterns. The distribution and type of color cells may change throughout the snake's life so that juveniles look very different from adults. The best-known examples of these are the Emerald Tree Boa, *Corallus caninus*, and the Green Tree Python, *Morelia viridis*, in both of which juveniles are bright yellow (rarely red) but change to green by the time they are about one year old. The convergent evolution of these two species, from opposite sides of the world, is discussed in more detail on page 52.

← The South American Rainbow Boa, *Epicrates cenchria*, is so named because of the iridescent nature of its scales, an example of interference coloration.

→ (Top) The brightly colored South American Coral Snake, *Micrurus lemniscatus*, is an example of warning, or aposematic, coloration.
(Middle) Baron's Green Racer, *Philodryas baroni*, is a handsome, fast-moving, diurnal species from the drier parts of Argentina, Bolivia, and Paraguay.
(Bottom) A particularly colorful specimen of the Australian Carpet Python, *Morelia spilota*, belonging to the subspecies *cheynei*, sometimes known as a Jungle Carpet Python.

Ornamentation

Snakes lack the ornamentation we see in many birds and lizards, as visual display is not an important part of their lives. Flaps and protuberances, the purposes of some of which are unknown at present, are mostly confined to the head and are found in only a few species.

A single modified, thorn-like scale is found over each eye in such species as the Desert Horned Viper, *Cerastes cerastes*, or there may be a small cluster of such scales, as in the Eyelash Pit Viper, *Bothriechis nigroadspersus*, and the Many-horned Viper, *Bitis cornuta*. The Sidewinder, *Crotalus cerastes*, and three species of false-horned vipers, *Pseudocerastes*, have a structure consisting of raised and enlarged scales over each eye. The function of these is not known, but may serve to disguise the outline of the snake's head or its eyes.

Snakes with protuberances on their snouts are more common and widespread. Examples include the three species of Madagascan leaf-nosed snakes, *Langaha*, and the Rhinoceros Ratsnake, *Gonyosoma boulengeri* (until recently known as *Rhynchophis boulengeri*). The Nose-horned or Sand Viper, *Vipera ammodytes*, is a European species with an upturned snout, and another viper, the Rhinoceros Viper, *Bitis nasicornis*, has a cluster of pointed scales on its snout. The purpose of these nasal structures is unclear, but in one species—the aquatic Tentacled Snake, *Erpeton tentaculatum*—the paired appendages contain nerves that can detect vibrations and changes in pressure in the water around them, enabling the snake to find its fish prey in turbid water or at night.

← The Desert Horned Viper, *Cerastes cerastes*, does not always have horns; some lack horns, and even the same clutch of eggs can produce horned and non-horned individuals.

→ A number of snakes have horns comprised of a single, large, pointed scale, especially pronounced in this spectacular and well-named Rhinoceros Viper, *Bitis nasicornis*, from Africa.

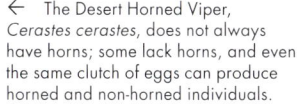

The sensory world of snakes

An essential element in the way organisms live involves the gathering of information about their surroundings. Because of their evolutionary history as burrowing animals, snakes use their senses and sense organs in ways that are different to most other animals, including ourselves.

Most snakes' eyesight is generally poor compared with humans, as is their sense of hearing (although they are not deaf, as is popularly believed). To compensate, they have an acute sense of smell, and many species have a remarkable ability to detect minute temperature changes.

CHEMICAL COMMUNICATION

Chemical communication (smell) is very important. When a snake detects some change in its environment it immediately flicks out its tongue through a notch in the upper jaw, known as the lingual fossa, and uses it to pick up scent molecules. These are returned to the mouth, where the twin tips of the forked tongue are inserted into a pair of sacs on the roof of the mouth, known as the Jacobson's organ. The scent molecules are transferred from here to

Jacobson's organ

Snakes use their forked tongue to pick up scent particles, before transferring them to the Jacobson's organ, situated in the roof of the mouth.

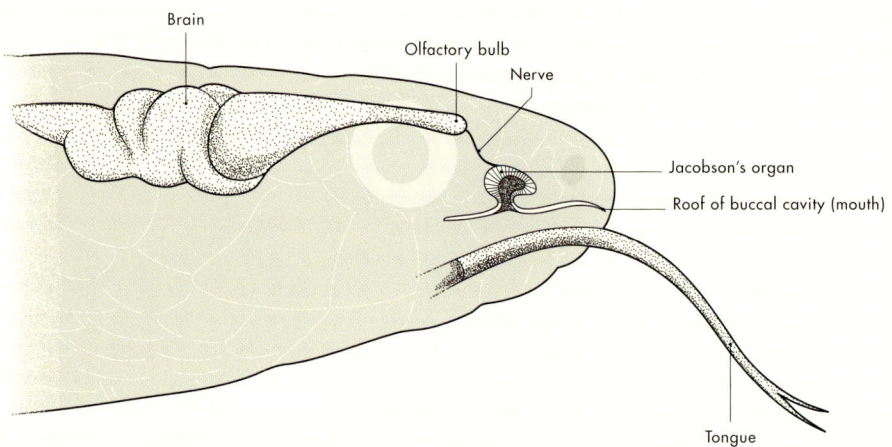

Brain

Olfactory bulb

Nerve

Jacobson's organ

Roof of buccal cavity (mouth)

Tongue

← A Brown House Snake, *Boaedon capensis*, extends its tongue to pick up scent molecules in the air, before transferring them to its Jacobson's organ, where they are analyzed and passed to the brain.

↓ A Lichen-colored Snail Sucker, *Sibon longifrenis*, flicks out its forked tongue in response to the photographer's presence. Scent is the most important sense to many species.

the olfactory part of the brain, along with messages obtained through the nostrils, and analyzed.

Snakes regularly monitor their surroundings by tongue-flickering. They use the information obtained to hunt for prey, detect potential enemies, and search for mates. Their olfactory system is extremely sensitive, and snakes have been known to sit by rodent trails that are many days old in wait for prey to pass by.

As snake-keepers will no doubt confirm, captive snakes become agitated if food is placed in the same room, even if it is some distance from the cages and inside a container, and this is often accompanied by an increase in the frequency of tongue-flickering.

SIGHT

Unusually for predators, most snakes have poor eyesight. The most primitive snakes, the Scolecophidia, all of which are confirmed burrowers, have only rudimentary eyes, covered with one or more scales, and capable only of distinguishing light from dark. The more advanced snakes have, to some extent, reinvented the eye, but only to a limited degree.

Focusing is very primitive compared with other vertebrates, and consists, for the most part, of moving the lens backward and forward. In addition, the cells lining the retina—the rods and cones—are not as well organized as they are in other vertebrates, and some species lack one or the other of them altogether. All these drawbacks result in the inability to spot and identify stationary objects, although they are more alert to moving objects.

Snakes that hunt by day, such as the various racers, whipsnakes, and garter snakes, tend to have large eyes with circular pupils. Nocturnal hunters also have large eyes, but their pupils usually take the form of vertical ellipses, which can be closed down to narrow slits in bright light. Horizontally elliptical pupils occur only in the Asian vine snakes, *Ahaetulla*, of which there are 21 species, and the three African twig snakes, *Thelotornis*. These horizontal or "wrap-around" pupils allow the snake to see forward with both eyes (binocular vision) and thus to judge distance accurately. In the *Ahaetulla*

↖↖　The pupils of the Asian vine snakes, *Ahaetulla* species, are horizontally elliptical, allowing them to look forward along the grooved snout and judge distances accurately.

↑　The vertical pupil in this strictly nocturnal tree snake, *Lycodryas pseudogranuliceps*, from Madagascar, has closed down to a small aperture.

↗　The diurnal Parrot Snake, *Leptophis ahaetulla*, from South America, has large round pupils. It hunts lizards, frogs, and invertebrates, largely by sight.

↗↗　The eyes of the Aquatic Coral Snake, *Micrurus surinamensis*, which feeds mostly on eels and other fish, are situated toward the top of its head so that it can look upward when it is partially submerged.

species, this ability is further enhanced by a long narrow snout with concave sides along which the snake can sight its prey.

Snakes' eyes are usually positioned on the sides of their heads, giving them a wide arc of vision to the front and back, useful for locating prey and predators. A few species have their eyes positioned on top of their heads, looking straight up. These are either aquatic snakes that rest on the water surface, such as the file snakes, *Acrochordus*, and the unusual aquatic coral snake, *Micrurus surinamensis*, or species that burrow in sand leaving just their heads exposed, such as the Arabian Sand Boa, *Eryx jayakari*, and the Namib Side-winding Adder or Péringuey's Adder, *Bitis peringueyi*.

HEAT DETECTION

Heat detection is highly developed in several groups of snakes, and has evolved more than once. It is unique to snakes and, in the species that have it, more than compensates for poor sight and hearing.

Specialized heat-sensitive pits are lined with epithelial cells that are connected to the brain by nerves. Amongst the boas and the pythons, pits are located in or between the scales bordering the mouth. Not all species have them, though, and where present they may be large and numerous or small and few. They are most noticeable in the larger, mammal-eating species belonging to the genera *Corallus*, *Sanzinia*, *Liasis*, *Morelia*, *Python*, and *Malayopython* but are completely absent in other species such as the Common Boa, *Boa constrictor*, and all of the Erycinae (sand boas, rosy boas, and rubber boas).

The most highly developed heat pits are found in the pit vipers (subfamily Crotalinae), a group that includes rattlesnakes, *Crotalus* and *Sistrurus*, and several other genera in the New and Old World, such as *Bothrops* and *Trimeresurus*. In these snakes the pits are paired, and located on either side of the head, just below an imaginary line between the eyes and the nostrils. They are directed forward and look like an extra pair of nostrils. Some pit vipers are known as *cuatro narices* or "four nostrils" in parts of Latin America.

These pits are more sophisticated than those of the boas and pythons, having two chambers separated by a membrane. The ambient temperature is detected by the inner chamber, while the outer chamber detects heat originating from a warm-blooded animal.

In experiments, some pit vipers have been shown to detect temperature differences of as little as 0.001 °C. Because the organs are paired, they work together "in stereo" to assess range as well as direction, and the snake is able to accurately strike at prey, even in total darkness, usually targeting the head or the thoracic region where the venom will act most quickly.

HEARING AND OTHER SENSES

Snakes' hearing is not on a par with most other vertebrates, but they can pick up vibrations through the ground, via their lower jaws, which are in contact with the quadrate bone and the columella, a small bone that transmits vibrations to the inner ear (equivalent to the stapes in mammals). Low-frequency airborne sounds may also trigger vibrations, but, because they have no eardrum, snakes can only detect a limited range of airborne vibrations, much less than mammals.

← Seen from this angle, the heat pits of the Jumping Viper, *Metlapilcoatlus nummifer*, are clearly directed forward.

↗ As in all pit vipers, the heat-sensitive pits of this White-lipped Island Pit Viper, *Trimeresurus insularis*, are positioned roughly between the eyes and the nostrils.

→ In the Emerald Tree Boa, *Corallus batesii*, the heat pits are positioned between the upper and lower labial scales.

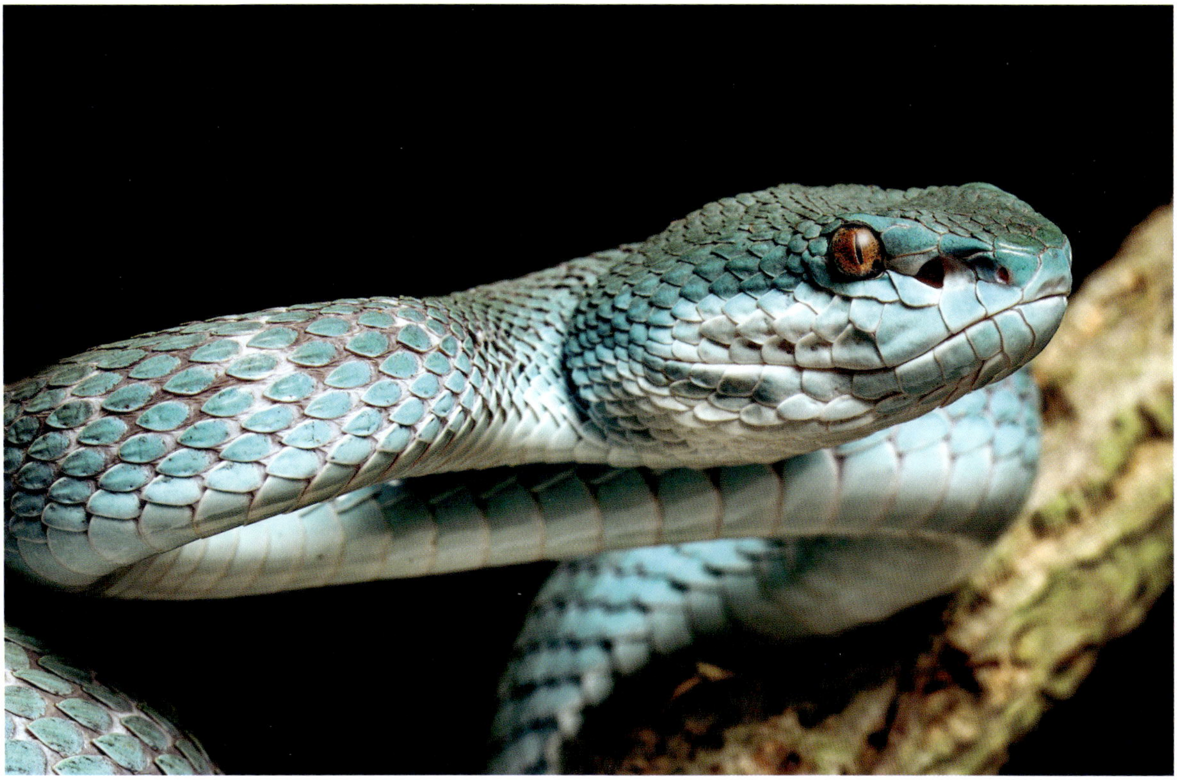

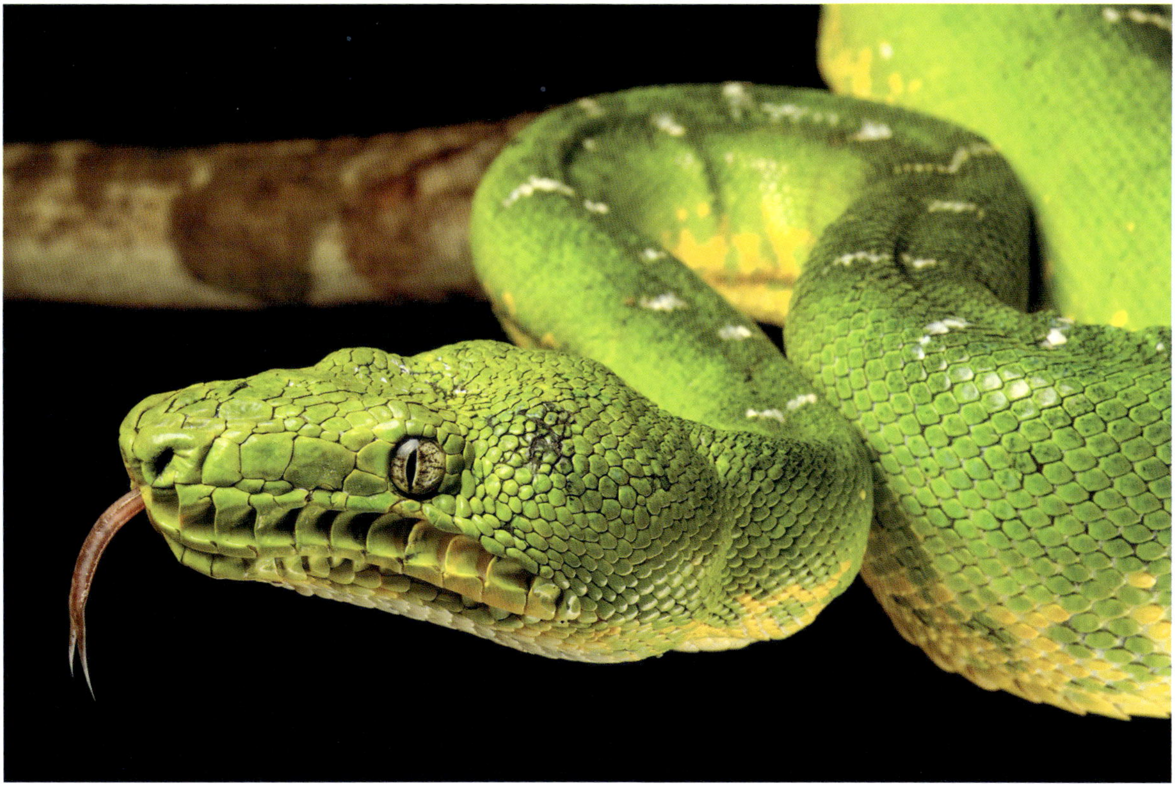

SENSE ORGANS IN THE SCALES

A number of structures on and between the scales of snakes are thought to carry information about their surroundings to the nervous system. The most obvious of these are apical pits, paired sense organs situated either side of the keel or midline of each dorsal scale, absent in some species but present in most. The functions of these pits, and others like them that are found on the head scales, are uncertain, but they are served by nerve endings and probably relay information about the immediate environment to the snake's brain. They may be sensitive to light, temperature, touch, or airborne chemical stimuli, or possibly a combination of these.

Similarly, there are small tubercles on the scales around the heads of most snakes, sometimes numerous but at other times very sparsely distributed. They are thought to be organs of touch, and, in some species, especially aquatic snakes, the tubercles are equipped with a small bristle, which may detect movements in the water caused by nearby animals.

The only such structures that have been investigated thoroughly are those in the paired appendages on the snout of the Tentacled Snake, *Erpeton tentaculatum*, from Southeast Asia. This aquatic species feeds on fish and hunts in turbid water and at night. The sense organs in its tentacles consist of clusters of nerve endings that apparently react to vibrations in the water caused by the movement of nearby fish, enabling it to strike rapidly and accurately even when the prey is not visible (see page 188).

↗ The paired apical pits in each scale show up well in this shed skin from a European Aesculapian Snake, *Zamenis longissimus*.

→ The unique paired structures on the snout of the Tentacled Snake, *Erpeton tentaculatum*, contain nerve endings that sense movements in the water nearby, helping it to detect prey even when it cannot see it.

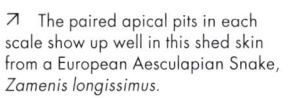

Convergent evolution

Animals that live in similar habitats often share certain aspects of their appearance and behavior, because each species has arrived at the same solution to the challenges they have in common. This is known as convergent evolution, and there are several well-known examples among the snakes.

The emerald tree boas, *Corallus caninus* and *C. batesii*, and the Green Tree Python, *Morelia viridis*, though belonging to different families and hailing from opposite sides of the world, are remarkably similar in their appearance, their behavior, and the way in which their color changes from yellow to green during their early growth. All three species live in rainforest habitats and feed largely on birds, relying on their camouflage to ambush their prey and their long curved teeth to grip and hold on.

Another example can be seen in the sidewinding vipers, *Crotalus cerastes* from North America (page 78) and *Bitis peringueyi* from southern Africa (page 100), and members of a third genus, the horned vipers, *Cerastes*, from North Africa (page 106). All of these live among loose, windblown sand dunes and, as well as looking similar and being of similar sizes, have evolved a similar method of locomotion (page 59).

The arboreal species *Sibon annulatus*, and its close relatives from tropical America, and the Blunt-headed

Slug Snake, *Aplopeltura boa*, along with other members of the Pareatidae from Southeast Asia, are all slug- and snail-eaters and share the characteristics of a large head, blunt snout, large eyes, and modified jaws adapted to extract snails from their shells.

On another level, the Australasian snakes belonging to the genus *Acanthophis*, commonly known as death adders, bear a striking resemblance to vipers, especially to members of the African genus *Bitis*, having stout bodies, a broad, triangular head, long fangs, heavily keeled scales (apart from one species), and sedentary habits. Despite their collective common name, they are not adders but members of the Elapidae. It is assumed that their body plan, hunting strategy, and reproductive habits have evolved to fill the niche left vacant by the absence of true vipers in that part of the world.

← The blunt snout, large eyes, and vertical pupils are characteristics of Asian snakes that feed on snails, such as this Montane Snail Snake, *Asthenodipsas vertebralis*, and are paralleled by South American species with similar feeding habits.

↗ The Many-horned Adder, *Bitis cornuta*, displays the short, heavy-bodied characteristics of an ambush predator. This species waits for prey to come within range before striking rapidly and forcefully.

→ Although the Australian death adders, *Acanthophis* species, belong to the Elapidae, they display all the characteristics of vipers, their sit-and-wait counterparts in the Old World, hence their somewhat misleading common name.

Moving around

Snakes clearly need to find a way to move around in their environment as they hunt for food, escape predators, find mates, and thermoregulate, and they must pursue all these activities despite lacking limbs. They use a number of techniques, some more specialized than others, and most snakes can use more than one means of locomotion.

The most widespread types of locomotion fall into three main categories: serpentine locomotion on the land (crawling) and in water (swimming); concertina locomotion, often used by burrowing snakes; and rectilinear locomotion (crawling in a straight line), used mostly by heavy-bodied snakes when moving slowly. Three additional methods are used by a handful of species that live in particular habitats: climbing, gliding, and sidewinding.

To a large extent, habitat determines which method a snake uses. They are not mutually exclusive, however, so an individual may switch from serpentine locomotion to concertina locomotion or straight-line crawling, for instance, according to its needs.

SERPENTINE LOCOMOTION

This is the most common form of locomotion. The snake wriggles from side to side, using the sides of its body to push against irregularities in the ground, rocks, or vegetation. At any given time, several points along the snake's body are pushing simultaneously against a number of different fixed objects. As the snake moves, new parts of the body come into contact with the same objects, and so all parts of the body follow the same line and the snake moves forward steadily and almost imperceptibly. The speed of movement will depend on the roughness or smoothness of the substrate, the shape of the snake, and its reason for moving.

←　A South American Puffing Snake, *Spilotes sulphureus*, travels across rough ground using serpentine locomotion.

During swimming in open water the same movements occur but the body of the snake pushes against the resistance of the water. Snakes that spend a large part of their time in water may have heavily keeled scales, and a laterally flattened body shape (flattened from side to side), to improve their purchase on the water. Marine snakes have an oar-shaped tail.

Snake locomotion

Snakes use a variety of methods of locomotion depending on the situation in which they find themselves. Serpentine is the most common; the other methods are used occasionally, depending on species and substrate.

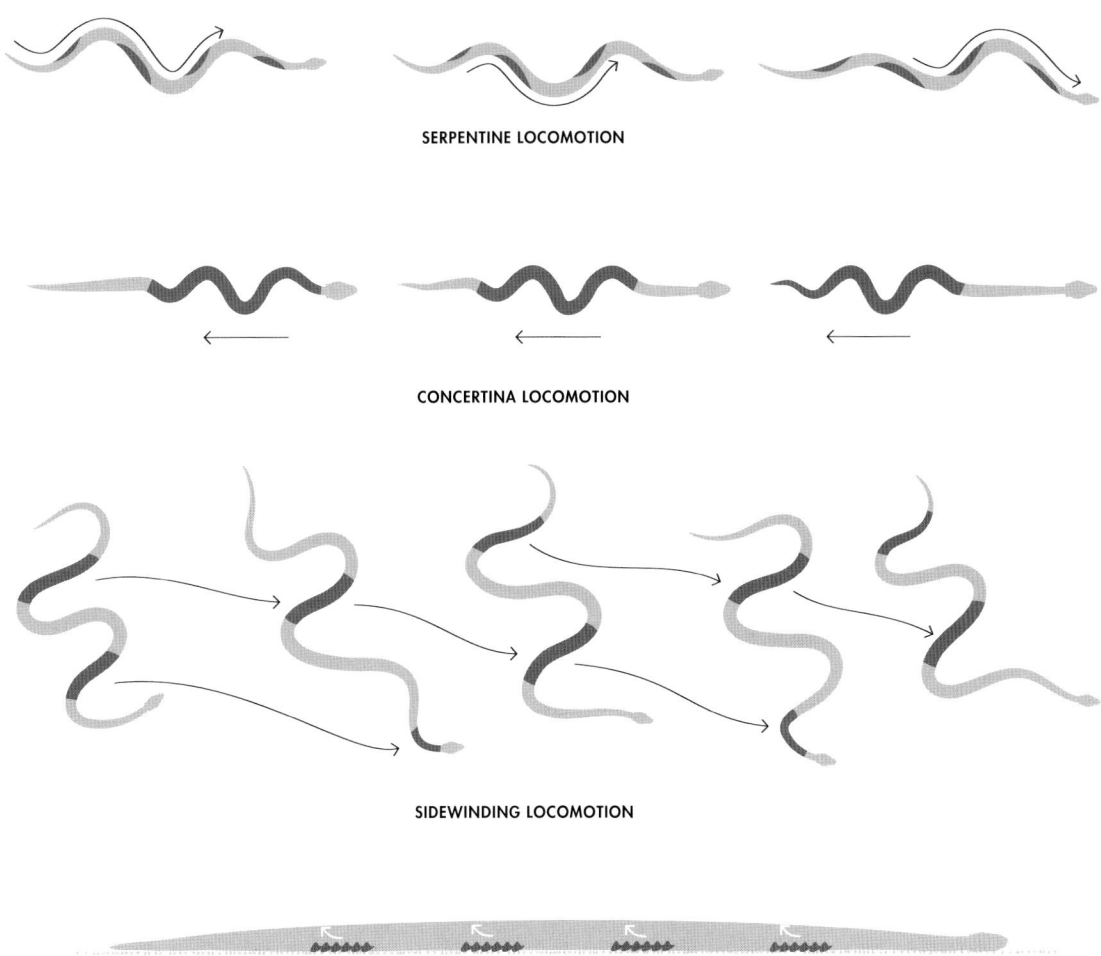

SERPENTINE LOCOMOTION

CONCERTINA LOCOMOTION

SIDEWINDING LOCOMOTION

RECTILINEAR LOCOMOTION

CONCERTINA LOCOMOTION

This is most often seen in burrowing species, but also when any snake is moving through narrow places, such as between rocks. A cycle of movement starts with the snake using the rear half of its body to jam itself against the sides of the burrow while it extends the front half. Once this is fully extended, it in turn is jammed against the tunnel walls while the rear part is pulled toward it. This results in a stop–start routine as the snake inches forward, in contrast to the more fluid action seen in other types of locomotion.

Most burrowing snakes are cylindrical in cross–section and have smooth or lightly keeled scales. They expand their ribs to increase their diameter when jamming themselves against the sides of the burrow, and may also kink their body to gain extra purchase. Shield-tailed snakes, Uropeltidae, are able to bend the vertebral column into a series of curves independently of the sides of the body, which remain parallel. The body thus becomes shorter and thicker so that the snake can wedge itself between the walls of the burrow with one part of its body while another part is thrust forward or drawn up. This is not a speedy method of moving, but burrowing snakes rarely need to move quickly.

↑   The Black Shield-tailed Snake, *Uropeltis melanogaster*, is a burrowing species that rarely emerges onto the surface. Like the other members of its family, it moves through its burrows by means of concertina locomotion.

↗   Many climbing snakes have ridges running along the edges of their ventral scales on either side; these help them grip rough surfaces when climbing.

→   Snakes' ventral scales are broad and overlapping on their trailing edge. There movement is controlled by muscles, enabling the snake to crawl forward.

RECTILINEAR LOCOMOTION

Heavy-bodied snakes, such as certain boas, pythons, and vipers, use the raised edges of their ventral scales to hook over irregularities and pull themselves forward, thus moving in a straight line. During the process, different sections of the body will be stretching forward, while other sections are pulling, as their muscles contract and relax in a series of waves from head to tail, resulting in a smooth progression forward. This is a relatively slow method of moving but presumably uses up less energy than serpentine locomotion. Rectilinear locomotion is also used by snakes in the final stages of stalking prey, as they inch forward slowly, and almost imperceptibly, to get within striking distance.

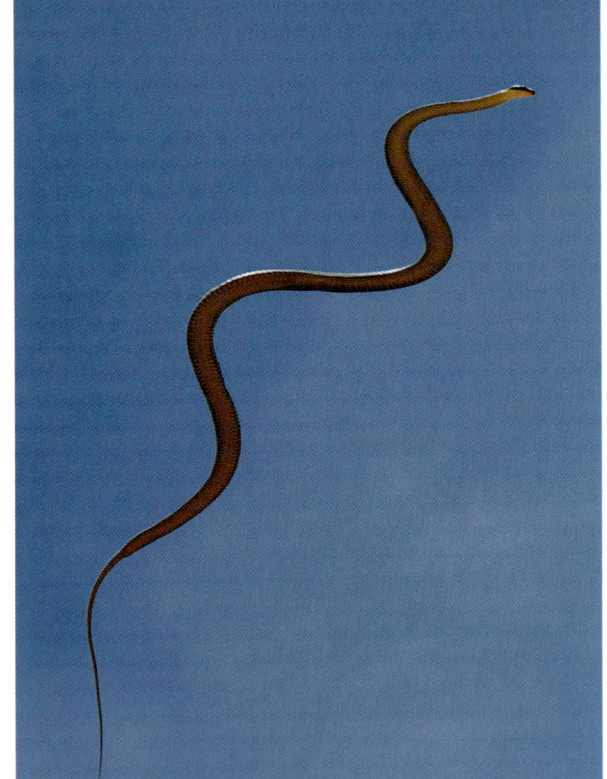

CLIMBING

Snakes that habitually climb amongst branches tend to have elongated bodies, and many are deep-bodied. This creates a girder-like cross-section, allowing the snake to cantilever the front part of its body across a wide gap while reaching for its next point of contact.

Other species, those that are not so specialized, move up tree trunks by using crevices in the bark to act as purchase points. They move in much the same way as they would when covering uneven ground, using serpentine locomotion, but in a vertical rather than a horizontal plane. A number of snakes from different genera have sharp keels at the edges of their ventral scales, where they meet the dorsal ones, creating a pair of ridges running along the length of the snake and providing it with a better grip on bark or other rough surfaces.

GLIDING

Gliding is a seemingly unlikely form of locomotion for snakes, lacking as they do the limbs or wings that other animals use to fly or glide. But snakes of the South and Southeast Asian genus *Chrysopelea* (the five species of which are known, rather inaccurately, as "flying" snakes) can launch themselves from tall trees and descend in a gradual and controlled manner to reach lower branches, or the ground, in order to escape predators. They do this by flattening their bodies until the ventral surface is slightly concave, to increase air resistance, while moving their bodies in a slowed-down sinuous serpentine motion. In this way they are able to steer and to maintain a position almost parallel to the ground throughout the "flight."

↖ A Leopard Snake, *Zamenis situla*, climbs a tree trunk by using the ridges on its ventral surface to gain purchase.

↑ A flying snake, *Chrysopelea* species, glides by flattening its body and "swims" through the air by forming a series of sinuous loops.

→ A Péringuey's Viper, *Bitis peringueyi*, moves rapidly across loose sand in the Namib Desert by means of a characteristic sidewinding technique.

SIDEWINDING

Snakes that live in deserts consisting of windblown sand dunes have difficulty in gaining purchase on the loose substrate, and their locomotion is impaired. Some species—those that live almost exclusively in this type of habitat—have evolved a specialized, and very effective, method of locomotion called sidewinding.

Starting from a resting position, they raise their head and neck off the ground and move it sideways, with the rest of the body providing an anchor-point. As soon as the head and body are back on the ground they, in turn, act as anchor-points and the rest of the body follows. In reality, this is a fluid movement, with one cycle beginning before the previous one has been completed, giving the impression that the snake is gliding sideways across the surface. The snake moves at about 45 degrees to the line of its body and leaves a trail of disconnected J-shaped impressions of its body in the sand.

The best-known sidewinding species are the Sidewinder, *Crotalus cerastes*, from the deserts of the North American Southwest, and the Namib Side-winding Adder or Péringuey's Adder, *Bitis peringueyi*. There are other sidewinders, such as the Saw-scaled Viper, *Echis carinatus*, and the horned vipers, *Cerastes* species, all from North Africa and the Middle East. A small number of snakes that live in coastal habitats, such as the Puff-faced Water Snake, *Homalopsis buccata*, and related species, may also use a similar sidewinding technique to cross mudflats, and some other snakes may switch temporarily to a primitive form of sidewinding when faced with a substrate that does not provide enough purchase for their normal mode of locomotion.

↑   An Ornate Tree Snake, *Chrysopelea ornata*, one of the "flying" snakes, is at home amongst high branches in rainforests. It will launch itself into the air if necessary to avoid predators.

↗   The Green Bush Viper, *Atheris chlorechis*, is a highly arboreal venomous snake from West Africa. All the *Atheris* species are covered in heavily keeled scales.

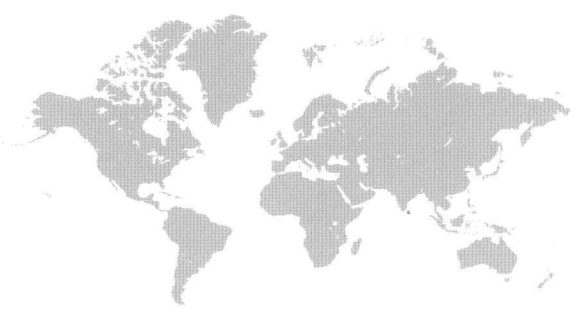

AHAETULLA NASUTA

Long-nosed Vine Snake

The snake with wrap-around eyes

SCIENTIFIC NAME	*Ahaetulla nasuta* (Lacépède, 1789)
FAMILY	Colubridae
SUBFAMILY	Ahaetuliinae
SIZE	3¼–5 ft (1–1.5 m)
REPRODUCTION	Viviparous, with 5–15 young
HABITAT	Forests

Various populations of vine snakes have recently been recognized as "cryptic species," in other words, species that are identical in appearance but show differences when subjected to genetic sequencing. Analysis of their DNA has shown that only the Sri Lankan populations should be assigned to this species, all the others being genetically distinct but morphologically the same: "hiding in plain sight," as some scientists have described it.

This snake, along with its close relatives, is remarkable for its long, pointed snout, large eyes, and horizontal pupils. These characteristics complement each other, and enable the snake to focus on a subject immediately in front of it by looking along its concave snout with both eyes. This binocular vision helps it to judge distances between the branches through which it moves, and a large proportion of its elongated body can be cantilevered out to bridge gaps.

Binocular vision also enables it to strike accurately at prey, usually arboreal lizards. Prey is killed by venom introduced via long fangs in the rear of its mouth, and it will hold the prey until it stops struggling, an essential strategy for snakes that hunt above the ground.

The Long-nosed Vine Snake is diurnal and highly arboreal, rarely descending to the ground and relying on its elongated shape and green coloration to provide excellent camouflage. Individuals tend to "freeze" when disturbed, enhancing their crypsis. If threatened, the snake will open its mouth widely and flatten its neck to display black and white markings. Bites can be painful but the venom is not dangerous to humans.

→ The long pointed snout and the horizontal pupils make the tree or vine snakes, *Ahaetulla*, of which there are 21 species in all, instantly recognizable.

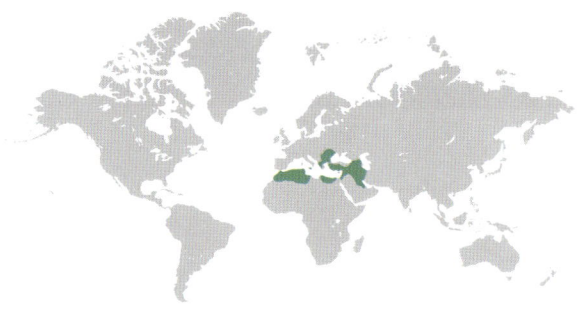

Javelin Sand Boa

Europe's only boa

SCIENTIFIC NAME	*Eryx jaculus* (Linnaeus, 1758)
FAMILY	Boidae
SUBFAMILY	Erycinae
SIZE	12–20 in (30–50 cm), occasionally longer
REPRODUCTION	Viviparous
HABITAT	Dry rocky, sandy, and scrub-covered fields, valleys, and hillsides

This small boa is the only European representative of the Boidae, a family that is more usually associated with tropical and subtropical regions. It only occurs in the warmer parts of the continent, along the eastern Mediterranean coastal region, including many small islands. The rest of its range falls within North Africa and the Middle East.

Like all sand boas, this is a secretive burrowing species, spending most of its time underground in rodent burrows or beneath flat rocks, where some heat penetrates through to the substrate. Its small, smooth scales and cylindrical cross-section are adaptations to a burrowing lifestyle, allowing it to move easily through the ground. When beneath the surface it moves by means of concertina locomotion (see page 56), inching its way slowly through tunnels or cracks or forcing its body through loose soil.

It probably finds most of its prey in this manner, feeding on rodents, although the snakes on some islands are much smaller than those on the mainland, and probably feed on small lizards and lizard eggs.

Although not endangered, owing to its large geographical range, some populations have been reduced or eliminated altogether through habitat destruction, mainly agricultural development and overgrazing. On a more positive note, in 2015 the presence of this species was confirmed from a small area in the south of Sicily, the first completely new reptile species for Italy for many years and a huge range extension from the rest of the European populations. In fact, individuals on Sicily resemble those in North Africa more than they do those in eastern Europe, leading to speculation that they may be more closely related to these populations.

→ Perfectly adapted to burrowing, with its cylindrical body, smooth shiny scales, and underslung jaw, the Javelin Sand Boa is the only member of the Boidae family that can be found in parts of Europe.

Reticulated Python

Jungle giant

SCIENTIFIC NAME	*Malayopython reticulatus* (Schneider, 1801)
FAMILY	Pythonidae
SIZE	16–20 ft (5–6 m), potentially to 30 ft (9 m) or more
REPRODUCTION	Oviparous
HABITAT	Rainforest, often around human habitation

One of the two largest snakes in the world, and probably the longest, even a half-grown Reticulated Python is a formidable predator. They have been known to take monkeys, crocodiles, and tigers as well as livestock such as goats, dogs, cats, and chickens. Humans have also been eaten on rare occasions.

Reticulated Pythons hunt mainly in the evening and during the night, actively seeking prey, which is detected by the heat-sensitive pits in the scales bordering the mouth (the labial scales). They are powerful constrictors, wrapping several coils around their prey and exerting pressure until it suffocates; bones may be broken in the process. They are slow-moving, crawling in a straight line (rectilinear locomotion) when on the ground, and often come to a complete halt when first encountered, rather than trying to escape.

Young and sub-adult Reticulated Pythons are partially arboreal, and typically rest in the branches of trees overhanging rivers, dropping into the water to escape if they feel threatened. They are powerful swimmers.

They are among the most prolific of snakes, laying clutches of 50–100 eggs, although 20–50 is more normal. The female coils around her clutch for the duration of the incubation period, although they appear to be unable to raise the temperature by muscular contractions, unlike some other large pythons.

Due to their large size and intricate pattern, Reticulated Pythons are one of the most desirable snakes in the skin trade; a total of 300,000–450,000 skins are traded each year, the majority of which come from Indonesia. They are also popular amongst zoos and amateur snake-keepers, and a number of strains, of a wide array of colors and markings, have been selectively bred to satisfy this market.

→ The Reticulated Python is a rainforest species that readily adapts to human-altered environments, such as plantations and villages, where its preferred prey of rodents and larger mammals is plentiful.

GONYOSOMA BOULENGERI

Rhinoceros Ratsnake

Long-nosed climber

SCIENTIFIC NAME	*Gonyosoma boulengeri* (Mocquard, 1897)
FAMILY	Colubridae
SUBFAMILY	Colubrinae
SIZE	3–5 ft (90–150 cm)
REPRODUCTION	Oviparous
HABITAT	Forests, including degraded forests, often near water

The purpose of the nasal protuberance that gives this species its name is unknown. It is present in both sexes and is covered in scales and is flexible. It may simply serve to break up the outline of the snake.

This is an arboreal snake, moving easily and rapidly through the branches of trees and understory shrubs, and most often found in humid river valleys at moderate altitudes, up to 5,000 ft (1,500 m). It lays clutches of 4–15 eggs which hatch after about two months. The young measure about 1 ft (30 cm) in length when they hatch and have the "rhino horn" right from the start. They are brown initially, changing to dark gray and eventually green or bluish green as they mature.

They feed on small vertebrates, lizards probably forming their main prey, although they also eat amphibians, small mammals, and birds. Populations appear to be reasonably secure, as much of their range is within protected areas, but habitat degradation is a problem in some places and small numbers are collected for the pet trade.

There are seven other species of *Gonyosoma*, all from Southeast Asia, mostly arboreal and green in color, though there are some exceptions. All of these other species lack the nasal appendage apart from *Gonyosoma hainanense*, described as recently as 2021 from Hainan, China; it is almost identical to *G. boulengeri*, except for small differences in color and the pattern of scales on its head.

The specific name *boulengeri* honors the Belgian-British herpetologist and polymath G. A. Boulenger, who worked at the Natural History Department of the British Museum from 1880 to 1920, describing 872 new species of reptiles as well as many fishes and amphibians.

→ The appendage growing from the snout of the well-named Rhinoceros Ratsnake is distinctive, but its function is a mystery.

Central American Eyelash Pit Viper

Multi-colored jungle dweller

SCIENTIFIC NAME	*Bothriechis nigroadspersus* (Steindachner, 1870)
FAMILY	Viperidae
SUBFAMILY	Crotalinae
SIZE	2–2½ ft (60–80 cm)
REPRODUCTION	Viviparous
HABITAT	Lowland tropical forest

This is a highly arboreal pit viper, usually found coiled a short distance above the ground, often in low-growing palms. For this reason members of the genus *Bothriechis* are known collectively as palm pit vipers. This species is remarkable for the pointed scales above its eyes, the "eyelashes" that give it its English name.

This pit viper occurs in many color forms, depending to some extent on its origin. The mottled coloration is the most common, but the shade of green and the extent of the markings vary. Other forms include tan, orange, or yellow; the bright yellow form, common in some parts of Costa Rica, is known as *oropel*, meaning tinsel or glitter. Polychromatic snakes such as these, in which there are a number of color and pattern types, are thought to benefit by confusing predators, and perhaps even prey, which are unable to build up a search image and so overlook them.

Eyelash Pit Vipers feed on a wide variety of prey, including small mammals, birds, and lizards. The juveniles often have colored tips to their tails, which they use as lures. There is some evidence that individuals "stake out" tropical flowers in order to ambush birds that are attracted to them for their nectar. Owing to their habit of resting at eye level, bites to humans are often on the face or hands. Bites are serious but rarely life-threatening in healthy humans.

The taxonomic relationships of this group of snakes has been revised. Many of the populations previously known as *Bothriechis schlegelii* have been reassigned to related species, and five new species have been described recently (2024). *B. schlegelii* is now restricted to the highlands of Colombia. Lowland populations from Central America are now mostly *B. nigroadspersus* but some have been reassigned to the species *B. supraciliaris*. Of the new species, only *B. schlegelii* and *B. supraciliaris* have the eyelash-like superciliary scales.

→ The Eyelash Pit Viper is an iconic species from Central America. It is fairly common, occurring in many colors and patterns that camouflage the snake against a variety of lichens, mosses, and other vegetation.

Green Bush Viper

A beautifully colored arboreal viper

SCIENTIFIC NAME	*Atheris chlorechis* (Pel, 1852)
FAMILY	Viperidae
SUBFAMILY	Viperinae
SIZE	18–24 in (45–60 cm)
REPRODUCTION	Viviparous
HABITAT	Rainforests

Also known as the Western Bush Viper, this is an arboreal viper that shows strong convergence in habits and appearance with arboreal vipers in Asia, *Trimeresurus* (page 76), and South America, such as *Bothriechis* species (page 70). Like those species, it is a well-camouflaged ambush predator that rests motionless until prey comes within range.

The body of this viper is slightly flattened from side to side, a characteristic shared by other arboreal snakes that enables it to bridge the gaps between branches, and it has a long, prehensile tail. By day it shelters among leaves, usually 3–6½ ft (1–2 m) above the ground. Its scales are heavily keeled, giving it a rough appearance, especially on the head; this disrupts its outline somewhat and, coupled with the bright green coloration, provides

good camouflage. The newborn young are tan at birth but change to a mottled green pattern within 24 hours.

Like other bush vipers, this species lives in tropical forests, especially along the edges of swamps and watercourses. Whereas other species of *Atheris* have suffered greatly from habitat destruction and persecution, and are classified as endangered, the Green Bush Viper seems to have adapted to life in secondary forests, banana plantations, and low-level agriculture,

Because of its habitat, and the thinly populated region in which it occurs, bites to humans are unusual. There is no antivenom for this species and existing antivenoms are not effective. Bites, however, are not thought to be serious, although they may diminish the ability of the blood to clot, resulting in heavy bleeding from the wound.

→ To some extent, the West African *Atheris* species, such as this Green Bush Viper, resemble the arboreal pit vipers, genus *Bothriechis*, from Latin America, both in appearance and in habits.

CHARINA BOTTAE

Rubber Boa

A tiny boa

SCIENTIFIC NAME	*Charina bottae* (Blainville, 1835)
FAMILY	Boidae
SUBFAMILY	Charininae
SIZE	1⅓–2½ ft (40–80 cm)
REPRODUCTION	Viviparous
HABITAT	Upland woodlands, grasslands, and desert edges

This is the world's smallest boa, rarely growing to more than 2 ft (60 cm) and often smaller. It is also the most northerly occurring boa. The smooth scales and pliable skin give rise to its common name. Rubber Boas are highly nocturnal and lead secretive lives, sheltering under rocks and logs by day. In the pine forests they also shelter under sections of fallen bark.

Rubber Boas are inoffensive and assume a characteristic posture if threatened: they coil their body into a ball with the head in the center. At the same time, the blunt tail may be raised above the coils and slowly waved around as though it is the head (page 203), and they sometimes make striking movements with it. Many individuals have scarred tails as a result, from wounds inflicted either by predators or by rodents defending their nests.

It feeds largely on nestling mice, shrews, and voles; small mammals make up two-thirds of its diet. The rest of its food consists of adult and nestling birds, lizards and their eggs, and salamanders. It constricts its prey before swallowing it head-first. When feeding on nestling rodents it usually eats the whole litter, and may be constricting one while it is swallowing another.

Females give birth in the autumn to 1–10 (average 4) pink or pale tan young measuring about 8 in (20 cm). It seems likely that females only breed every two or three years under normal circumstances, as they enter hibernation shortly after giving birth and do not emerge again until the following spring. They are long-lived, and captives have survived for over 20 years. A second species, *Charina umbricata*, is sometimes recognized but is more often considered to be a subspecies.

→ The diminutive Rubber Boa, often only 16 in (40 cm) or less in length, is nevertheless a typical member of the Boidae, and a powerful constrictor of small mammals and birds.

TRIMERESURUS INSULARIS

White-lipped Island Pit Viper

Green, blue, or yellow

SCIENTIFIC NAME	*Trimeresurus insularis* Kramer, 1977
FAMILY	Viperidae
SUBFAMILY	Crotalinae
SIZE	Up to 3¼ ft (1 m)
REPRODUCTION	Viviparous
HABITAT	Rainforest

This species, also known as the Sunda Pit Viper, occurs in a variety of colors, the most common of which is green, but spectacular blue and yellow forms are found on a number of islands within the Lesser Sunda chain.

The blue form probably owes its existence to a mutation in which the layer of xanthophores, oil-filled cells that produce a yellow filter, is missing. Under normal circumstances, this yellow filter overlying the blue scales would result in green coloration. The blue form is arguably one of the most beautiful snakes in the world.

The White-lipped Island Pit Viper is highly arboreal and strictly nocturnal. It typically hunts by hanging from a low branch, head-down, with its head drawn back, waiting for prey to pass beneath it. The edges of streams, where frogs and toads are abundant, are favorite hunting sites. It may also take up a more horizontal position facing rock walls or tree buttresses, waiting for climbing lizards such as geckos to come within striking distance. Individuals will return to the same place and position night after night until they are successful. Apart from amphibians and lizards, its diet also includes small mammals and snakes.

This is the most common snake on some of the islands on which it occurs, and individuals are often killed when they are found near human settlements; nevertheless, populations appear to be stable. Bites to local people are frequent and can be painful but are rarely, if ever, lethal. Small numbers are collected for the pet trade due to the varied and attractive color forms.

→ The White-lipped Island Pit Viper was only described as a distinct species in 1977 but has attracted a lot of attention due to its spectacular and varied color forms.

CROTALUS CERASTES

Sidewinder

A desert specialist

SCIENTIFIC NAME	*Crotalus cerastes* Hallowell, 1854
FAMILY	Viperidae
SUBFAMILY	Crotalinae
SIZE	20–30 in (50–80 cm)
REPRODUCTION	Viviparous
HABITAT	Deserts

The American Sidewinder is a small rattlesnake, famous for its looping form of locomotion, known as sidewinding, from which it gets its common name (page 59). The Sidewinder occurs in the sandy inland desert regions of the American Southwest, and is adapted to living on dunes, dry river washes, and areas of sparse vegetation, although it is sometimes found in gravelly or rocky habitats.

All the scales are keeled, those down the center of the back more so than the others. Its most distinctive feature is the pair of large scales above the eyes, the supraoculares, that are raised slightly and pointed, like blunt horns. There is some variation in overall coloration, so that the snakes match the different substrates on which they live.

Apart from early spring, when it is active during the day, it is a nocturnal species, often traveling several hundred yards in a single night and hiding by day in shallow crater-like scoops, which it makes by shuffling down into the sand until most of its body is thinly covered; windblown sand may further add to its cover. It often rests in the shade of a bush, a position from which it ambushes its prey, consisting of lizards and small mammals—which often make their burrows in similar situations. Its venom, which consists of various components, is effective against its relatively small prey. It is not considered to be dangerous to humans generally, although a few fatalities have occurred.

Sidewinders give birth to live young, with litters of 1–20, with an average of nine, born at the end of summer, usually in September. The newborn snakes feed on small lizards, many of which hatch around the same time of year. Unusually for a rattlesnake, females are slightly longer than males.

→ A Sidewinder (*Crotalus cerastes*) from the American southwest, tasting the air with its tongue.

Paradise Tree Snake

The "flying" snake

SCIENTIFIC NAME	*Chrysopelea paradisi* Boie, 1827
FAMILY	Colubridae
SUBFAMILY	Ahaetuliinae
SIZE	Up to 4¼ ft (1.3 m)
REPRODUCTION	Oviparous
HABITAT	Forests

One of several snakes in this genus popularly known as "flying" snakes. The Paradise Tree Snake is highly arboreal, rarely coming down to the ground. Its adaptations to this lifestyle include an elongated body, a narrow head, and an ability to glide from tall trees to lower vegetation as a means of escaping predators.

Many climbing snakes have ventral scales that are more-or-less flat with a sharp ridge at either side, where they meet the dorsal scales (see page 57). This produces a ridge along each side of the body that can engage with rough surfaces such as the bark of trees, enabling the snake to climb vertical surfaces easily and to crawl along horizontal branches. The so-called flying snakes, of which there are four others in addition to this one, have taken this a stage further: their undersides can be pulled upward, forming a concave surface from neck to tail. This increases air resistance so that when the snake launches itself from the forest canopy it can, by moving its body in a series of sinuous loops, "swim" through the air. The concave cross-section not only slows down its descent but, by tilting itself from side to side, the tree snake appears to be able to control the direction of its descent, at least to some extent.

Chrysopelea paradisi appears to eat mainly lizards, but also small mammals, birds, and frogs, but its natural history is poorly known. It subdues prey using venom delivered by elongated fangs at the back of its mouth. Though painful, bites are not dangerous to humans.

→ A particularly colorful example of the Paradise Tree Snake, in its forest habitat.

DEALING
WITH THE
ENVIRONMENT

Temperature and water balance

Environmental conditions play a vital role in the lives of snakes, as they do for all animals and plants. Their senses constantly monitor physical and biological variables, and the snakes react accordingly. Biological factors include interactions with other animals, which may be predators or prey, and with members of their own species, which may be potential mates or rivals. The most important physical factors are temperature and water balance.

TEMPERATURE CONTROL

Because they rely on outside sources of heat to maintain their bodies at sustainable temperatures, climatic conditions are among the most important factors in determining where snakes live, and how they live. Many details of their activity patterns reflect their environments.

As we have seen, snakes are more numerous, in both species and numbers, in warm places, especially in the tropics and subtropics. The ambient temperatures in these regions are close to snakes' preferred temperatures, and so they need to expend little or no energy shuttling between hot and cold places to maintain a viable temperature. Snakes that live in cold places spend much of their time attempting to raise their body temperature above that of the environment. On the other hand, snakes from very hot places have the opposite problem: how to prevent themselves from becoming overheated.

← A Cuban Boa, *Chilabothrus angulifer*, resting at the base of a buttress tree in rainforest. This is by far the largest snake in the Caribbean region, reaching 16 ft (5 m) or more in length.

↗ A tributary of the Kinabitangan River, Borneo. Reticulated Pythons, *Malayopython reticulatus*, and several arboreal snakes frequently rest in branches overhanging the water.

WATER BALANCE

Snakes have relatively impermeable skins and scales, so water loss is not so great as in some other animals, notably the amphibians. All snakes drink water when it is available, and they obtain some from their food.

Desert snakes, some of which never encounter standing water, have the greatest challenge. They employ a number of strategies to minimize water loss, including retreating underground and coiling to reduce their surface area. Some desert snakes use warning sounds such as tail rattling or scale-sawing rather than hissing, and it has been suggested that these behaviors avoid expelling air, and therefore water vapor, through the mouth.

Sea snakes obtain water from their food (mostly fish) and use specialized glands to remove excess salt, so maintaining the correct salt balance (see page 97).

Thermoregulation

For many snakes, especially those from temperate regions, the process of thermoregulation—keeping their body temperature within certain limits—determines their diurnal and seasonal activity patterns.

A PREFERRED TEMPERATURE RANGE

Contrary to popular belief, snakes and other reptiles are not "cold-blooded." They are ectotherms, which means that they rely on an external heat source (the sun, either directly or indirectly) to maintain the temperature of their blood, and the rest of their body. This depends on the snake behaving in a way that will keep its temperature within preferred values for as long as possible. From experiments and observations of wild snakes, it appears that a snake's preferred body temperature is about 85 °F (30 °C). This value is common to most species, regardless of where they live, although there is some variation.

At either end of the scale, snakes may still be active at temperatures as low as 50 °F (10 °C) and as high as 105 °F (40 °C). They need these "margins" so that they can move around to bask or shelter. A snake resting under a rock, for instance, may have a body temperature well below its preferred value, but it can still emerge to seek out a warmer spot. Once it has raised its body temperature in this way, it can go about its daily activities, such as searching for prey or a mate.

← A European Adder, *Vipera berus*, basking in the open, absorbing heat from a rock but not straying too far from cover.

↗ Asp Viper, *Vipera aspis*, a European species that occurs up to 11,000 ft (3,300 m) in the Swiss Alps, but which is equally at home in fields and forests at lower altitudes.

As the snake's body cools down, it reaches a point known as its critical minimum, where it loses the power of locomotion and cannot therefore move to somewhere warmer. This value varies with species, but it is usually a few degrees below 50 °F (10 °C) for those that have been studied. Before it reaches this stage, the snake would normally try to move to somewhere warmer. If the temperature continues to fall, it will eventually reach a lethal minimum temperature, and the snake will succumb. Similarly, there is a critical upper temperature that, if exceeded, will cause the snake to become immobile; if the temperature continues to rise, the snake will overheat and die.

Thermoregulation in diurnal snakes

Each snake species has a preferred temperature range that it tries to maintain. Either side of this are the critical temperatures and then the lethal temperatures. The critical maximum temperature is much closer to the snake's preferred temperature range than is the lethal minimum, which means that snakes are more vulnerable to overheating than they are to freezing. They avoid lethal temperatures at either end of the scale by a number of strategies, mostly behavioral. Understanding these is the key to appreciating how snakes live in extreme environments.

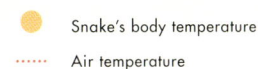

Snake's body temperature

.... Air temperature

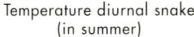

Temperature diurnal snake
(in summer)

85°F/30°C

0°

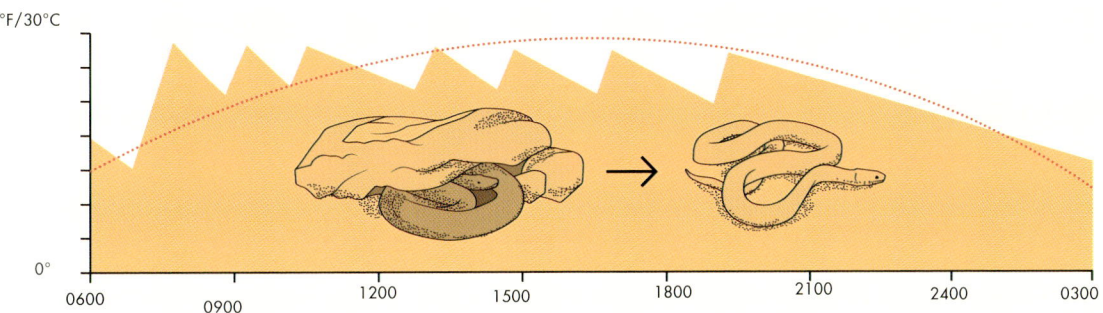

0600   0900   1200   1500   1800   2100   2400   0300

Time of day

COPING WITH THE COLD

Snakes can be found in many of the colder parts of the world where temperatures are far lower than their preferred body temperatures. The Adder, *Vipera berus*, for instance, occurs well into the Arctic Circle in northern Scandinavia and Russia. In North America the Red-sided Garter Snake, *Thamnophis sirtalis parietalis*, is found as far north as Hudson Bay in Canada, and in South America the Patagonian Lancehead, *Bothrops ammodytoides*, occurs at least as far south as 47°S on the Valdés Peninsula, in southeastern Argentina.

As well as latitude, altitude also affects temperature, and the highest recorded species include the Himalayan Pit Viper, *Protobothrops himalayanus*, and the Hot Springs

↑ An Eastern Diamondback Rattlesnake, *Crotalus adamanteus*, flattening its body against the sandy substrate to absorb as much heat as possible.

↗ An Eastern Garter Snake, *Thamnophis sirtalis*, basking among leaves on the forest floor.

→ Rainforest species, such as this *Dendrophidion nuchale* from Central America, rarely bask in the open.

Snake, *Thermophis baileyi*. Neither of these species, however, can maintain its preferred body temperature without adapting its behavior.

DAILY ACTIVITY PATTERNS

Behavior to cope with cold conditions can be divided into daily and seasonal patterns. On a daily scale, snakes from cold places bask, especially early in the day, to warm themselves up to a level at which they can operate efficiently. Many of these species are dark in color, thus optimizing heat absorption while they are exposed to the sun. Basking takes the form of stretching out the body, flattening it and tilting it toward the sun to expose as much of the surface area as possible. Basking can also take place underneath objects that are warmed by the sun, such as slabs of rock, mats of vegetation, and pieces of tin or plastic. This allows the snake to absorb heat while still remaining out of sight of possible predators.

Nocturnal species will bask when they first emerge in the evening by flattening themselves against a rock or a tarred road, both of which retain heat after the sun has gone down. Once they are warmed up, snakes can limit the rate of heat loss by coiling up and reducing their surface area. In some cases, several snakes may coil up together so that they "borrow" heat from one another.

BLACK-HEADED SNAKES

A number of snakes from various regions and families have black or dark heads. The black-headed snakes, *Tantilla* species, are obvious examples, but there are many others. These species can initiate basking by exposing just their head while keeping the rest of their body under cover. The brain and sense organs warm up first so the snake can be alert to danger before emerging completely.

→ Florida Black-headed, or Crowned, Snake, *Tantilla relicta*, a small, secretive species.

SEASONAL ACTIVITY PATTERNS

As well as a daily thermoregulatory regime, snakes that live in cold regions adopt seasonal activity patterns that help them cope with the conditions. In areas where there are large temperature swings through the year, such as central and northern Europe, North America, and parts of central Asia, snakes may be nocturnal during the warmest months, diurnal in spring and autumn, and inactive during the winter.

← The Western Dusky Rattlesnake, *Crotalus triseriatus*, is a highland species that relies on basking to raise its body temperature.

↗ The Hot Springs Snake, *Thermophis baileyi*, can only survive at high altitudes because it lives in the vicinity of hot springs.

→ Dark coloration and diurnal basking allow the Himalayan Pit Viper, *Protobothrops himalayanus*, to survive at very high elevations.

Species that live in the coldest places tend to give birth to live young; in other words, they are viviparous or ovo-viviparous (see page 132). This allows the females to bask when conditions are suitable, moving as the sun's position changes during the day to warm up their developing embryos. For this reason, pregnant females will invariably bask for longer periods than their male counterparts.

THE CHALLENGES OF ALTITUDE

Of all the snakes that have adapted well to cold climates, the vipers are the most numerous. They absorb and conserve heat well because of their stocky shape,

they tend to be dark in color, and they mostly have live young, all characteristics that lend themselves well to life in cold places, and this includes high elevations as well as high latitudes. The highest-occurring snakes on most continents are vipers: the Himalayan Pit Viper, *Protobothrops himalayanus*, and Strauch's Pit Viper, *Gloydius strauchi*, can be seen as high as 16,000 ft (4,900 m) in the Himalayas, and there are several other examples of high-altitude pit vipers in Asia. In North America, the Western Dusky Rattlesnake, *Crotalus triseriatus*, occurs up to 14,000 ft (4,300 m) in central Mexico, while in Africa the Berg Adder, *Bitis atropos*, is found up to 10,000 ft (3,000 m).

The only colubrid that occurs in high mountain regions is the Hot Springs Snake, *Thermophis baileyi*, which lives up to 14,500 ft (4,440 m) in Tibet but avoids the temperature problem by living near hot springs.

Snakes in deserts

There are deserts in North and South America, Africa, Asia, and Australia. In total, deserts and semi-deserts cover about 30 percent of the Earth's land surface. It is hardly surprising, therefore, that many snakes have adapted to a desert lifestyle.

Deserts vary in character from place to place. When we picture a desert, we may think first of the great sand seas of the Sahara, the Namib, or the Atacama, but there are also rocky deserts, as seen in the American Southwest, the southern African Karoo, and central Australia.

Not all deserts are constantly hot, but most of them experience extreme temperature fluctuations, from blistering heat during the day to freezing-cold nights. Despite these challenges, snakes are ideally suited to live in deserts. Because they are ectotherms and don't rely on their metabolism to provide heat, they can subsist on a fraction of the amount of food that a bird or mammal of similar size would require, and they can fast for long periods of time, several months in some cases. This is helpful in places such as deserts, where food is in short supply.

The lack or shortage of water is also less of a problem for snakes, as their scaly skins limit the amount of water loss, while their excretory systems produce waste in the form of a semi-solid uric acid that requires very little water to carry it out of the body.

← The shifting sand dunes of the Arabian Desert, and similar places in other parts of the world, present great challenges to snakes. Only a few highly specialized species can survive here.

↗ The Rosy Boa, *Lichanura trivirgata*, is a nocturnal desert species, often occurring on dry, rocky hillsides.

→ The Horned Adder, *Bitis caudalis*, is fairly common in the dry sandy or rocky deserts and semi-deserts of southern Africa. Its coloration varies according to the substrate on which it lives.

Furthermore, desert snakes are largely nocturnal—to avoid overheating—and their unique sensory systems, consisting of smell, a heightened ability to detect vibrations due to their contact with the ground, and, in some cases, heat-detecting organs, are ideally suited to hunting in darkness.

Large areas of shifting dunes are inhospitable even for snakes, and only a small number of highly specialized species, such as the sidewinding vipers of the Colorado and the Namib Deserts, *Crotalus cerastes* and *Bitis peringueyi*, can eke out a living there. In rocky and scrubby deserts, snakes belonging to several families, including the vipers and pit vipers, Viperidae, and the Colubridae and their allies, can be numerous. Some representatives of other families are also found in these habitats, including members of the Boidae, particularly the North American Rosy Boa, *Lichanura trivirgata*, and members of the Erycinae subfamily, which includes the sand boas. In Australia, where there are no vipers or boas, their place is taken by some pythons, notably the Woma, *Aspidites ramsayi*, and elapids, including the viper-like Desert Death Adder, *Acanthophis pyrrhus*.

Snakes on islands

Snakes arrive on islands by three means: they may already have been on the land when it became separated from the mainland due to rising sea levels, subsiding land levels, or erosion; they may have rafted there among debris from the mainland, perhaps following a storm; or they may have been introduced deliberately or accidentally by human agency (see page 268).

Being relatively poor at dispersal, snakes have not fared well on small islands, and many of the world's islands are devoid of snakes altogether. Where snakes do occur, speciation may have taken place due to the "founder effect"—where colonies have descended from a relatively small gene pool due to lack of subsequent recruitment from the mainland. There are many examples, but just two are described here.

The islands in the Gulf of California, also known as the Sea of Cortez, were originally part of the mainland. Rising sea levels created islands from hill and mountain tops, and the snakes that lived on them were isolated.

← The snakes living on the various Galapagos Islands, *Pseudalsophis* species, are all closely related. They are thought to have evolved from a single landfall, originating from a population in South America.

↑ The Night Snakes, *Hysiglena ochrorhynchus*, living on Cedros Island, off Baja California, are distinct from mainland populations and are sometimes given the subspecific name *H. o. baueri*.

↗ An adult Galapagos Racer, *Pseudalsophis biserialis*, at Punta Pitt, San Cristobal Island.

Over many generations they diverged from their ancestors on the mainland until they formed new species. For example, there are five rattlesnakes endemic to single islands in the Gulf, and several unique subspecies. The endemic Santa Catalina Rattlesnake, *Crotalus catalinensis*, has lost its rattle through the process of evolution (see page 114), while another species, the San Esteban Rattlesnake, *C. estebanensis*, has a reduced rattle that may be missing in some individuals.

The Galapagos Islands, which are volcanic and have never been joined to the mainland, are famous for the speciation of finches and giant tortoises. But there are nine species of racers, *Pseudalsophis*, living on separate islands in the archipelago that also demonstrate island speciation. The first extant species to have evolved is thought to be the Españolan Racer, *P. hoodensis*. Its ancestral species, which is now extinct, is assumed to have rafted from the South American mainland, where a close relative still lives. From Española the racers dispersed to some of the other islands, where they evolved into the cluster of species recognized today. Fittingly, one of the newly described species, from Isabela, Fernandina, and Tortuga, was named *Pseudalsophis darwini* in 2018.

Snakes in oceans

Oceans pose particular problems for snakes, but some have found ways to overcome them. The sea snakes are members of the Elapidae, and number 72 species. There are also snakes that live in brackish water or which venture into the sea occasionally; these include the file or wart snakes, *Acrochordus* species, the homalopsid water snakes, and a few colubrids.

Snakes that live entirely in an aquatic environment have limited opportunities to thermoregulate, so they all live in tropical or subtropical waters. Respiration imposes a limiting factor on the length of time sea snakes can remain submerged. Among other adaptations, their lungs are proportionally larger than those of land snakes, and the posterior section of the lung (the saccular lung) is thick-walled and muscular. This is thought to enable the snake to shunt oxygenated blood from the saccular lung to the vascular lung, where gaseous exchange takes place; the saccular lung acts as a kind of "reserve tank."

Marine-dwelling snakes must also overcome the problem of salt balance. In terrestrial snakes, as in other animals, the kidneys regulate salt by removing excess when necessary, but sea snakes live in an environment that has a higher salt concentration than their bodies. There will be a tendency for water to move out of their bodies through osmosis until the concentrations are balanced, which would lead to dehydration.

Furthermore, sea snakes and file snakes eat fish, which are salty, and so excess salt will accumulate in their systems, and this must be excreted. They do this by using a specialized gland situated under the tongue, known as the sublingual gland, whose function is to remove surplus salt from the blood. A duct leads from the gland to the sheath surrounding the tongue so that every time the snake extends its tongue it pushes the concentrated salty water out.

The homalopsid snakes that live in brackish water, and which feed on fish and crustaceans, have independently evolved a salt gland in the roof of their mouths, the premaxillary gland. The few colubrid snakes that enter brackish water have no specialized means of reducing the salt concentration, and live within their body's salt tolerance level by spending only short periods of time in salty water.

← Although it lives mostly in fresh and brackish water, the Arafura File Snake, *Acrochordus arafurae*, occasionally enters the sea.

↑ Stokes' Sea Snake, *Hydrophis stokesii*, is a wide-ranging and adaptable species occurring in a number of different marine habitats.

Snakes in urban habitats

There can be no doubt that the encroachment of the human species, and the disturbance and the pollution that goes with it, is detrimental to wildlife, including snakes. There are many examples of snake populations suffering as a result of human activities (see pages 246–259), but there are a few cases where snakes appear to have benefited, at least temporarily.

Human activities, whether they are in rural or urban surroundings, invariably bring with them an increase in the rodent population, and this provides a welcome source of food for some of the larger snakes. In some places, snakes are encouraged, or at least tolerated, for this reason, provided they are non-venomous. In southern Africa, for instance, Brown House Snakes,

Boaedon capensis, frequently live in or around houses, barns, and outbuildings. The same is true of a number of snakes, including pythons, in Asia and Australia. Cockroaches, moths, and other insects, lured in by artificial light, attract geckos and other lizards that in turn are eaten by snakes—a human-induced food chain.

← According to recent reports, climate change has forced snakes in Australia to enter dwellings to avoid excessive heat. This highly venomous Northern Brown Snake, *Pseudonaja nuchalis*, was discovered in an abandoned building and is seen here in its defensive posture.

→ A garter snake basking on a man-made structure.

Damaged and derelict buildings, old bridges, and other infrastructure in poor repair, even abandoned vehicles, can provide good habitat for snakes, especially in otherwise featureless and unsuitable places. Snake researchers and enthusiasts know the value of these places: sifting through piles of old building materials, searching in and under abandoned vehicles, and inspecting cracks and crevices in masonry can be productive ways of finding snakes. Often, refuse dumps provide good habitat for the same reason and can encourage concentrations of snakes that may be sparsely scattered elsewhere. Where development has resulted in the draining of wetlands, a variety of semi-aquatic snakes may be incidental beneficiaries of artificial ponds, canals, ditches, and flooded gravel pits, while wildlife gardens can provide islands of suitable habitat in otherwise sterile landscapes, even though they are rarely designed and maintained with snakes in mind.

Péringuey's Adder

A sand-dune specialist

SCIENTIFIC NAME	*Bitis peringueyi* (Boulenger, 1888)
FAMILY	Viperidae
SUBFAMILY	Viperinae
SIZE	8½–10 in (22–25 cm), occasionally to 13 in (32 cm)
REPRODUCTION	Viviparous
HABITAT	Windblown sand dunes

One of the smallest vipers in the world and unique in its lifestyle and adaptations, Péringuey's Adder, sometimes known as the Namib Side-winding Adder, lives only on the narrow coastal belt, among the shifting dunes of the Namib Desert, one of the oldest and driest deserts in the world.

A Péringuey's Adder has no home range, nor permanent retreat in which to hide during the day, but shuffles down into the sand leaving an indistinct outline of its body, which quickly disappears as the wind blows sand over it, leaving only its upwardly pointing eyes and the tip of its tail breaking the surface. It moves by sidewinding, leaving characteristic tracks similar to those of the other desert sidewinder, the rattlesnake *Crotalus cerastes* (page 78).

Péringuey's Adder is predominantly orange in color, but the tip of its tail may be black, and is used as a lure. Its main prey is the lizard *Meroles anchietae*, with which it shares its habitat. It obtains most of its water from its prey but can also sip drops of condensing fog from its scales, flattening its body to increase its surface area.

It gives birth to 4–10 young measuring about 5 in (12 cm) in length, but little is known of its natural history. It seems to be numerous in places, although habit destruction caused by off-road vehicles is a problem. Some are killed on the few tarred roads that cross its habitat, and small numbers are collected illegally for the pet trade.

Bites are very rare, because there is no permanent human population in the area it inhabits. The few bites known led to localized pain and swelling that disappeared after a few hours.

→ A Péringuey's Adder is a perfect color match for the sand on which it lives. The tail has a black tip, which the snake will wiggle and use as a lure to attract lizards.

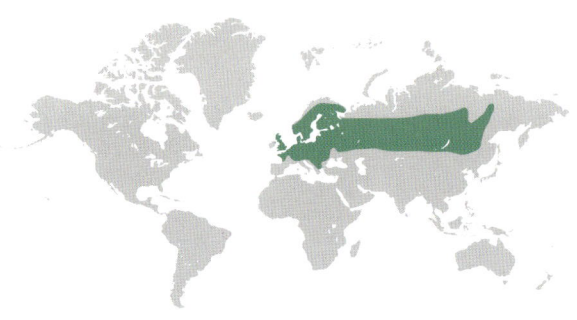

Adder

Most northerly snake

SCIENTIFIC NAME	*Vipera berus* (Linnaeus, 1758)
FAMILY	Viperidae
SUBFAMILY	Viperinae
SIZE	20–26 in (50–65 cm), occasionally up to 31 in (80 cm)
REPRODUCTION	Viviparous
HABITAT	Varied: heathland, grassland, marshes, and scrub

The Adder, or Northern Viper, has the most extensive distribution of any terrestrial snake, and occurs further north than any others, well into the Arctic Circle in Scandinavia and northern Russia.

This is one of the few snakes in which the sexes differ in coloration and contrast: males are usually pale silvery-gray with a dark gray or black zigzag, whereas females tend toward brown with darker markings. Occasional individuals of either sex may be uniform black, especially toward the north of its range.

Breeding takes place in the spring, immediately after hibernation, with males emerging first. When females emerge, rival males compete with each other for the opportunity to mate, raising the front halves of their bodies and intertwining them, while trying to push their rival to the ground. Evenly matched males may tussle for extended periods, but combats are usually over quickly. The victorious male usually mates with the female soon afterward.

Females usually breed every other year, with the "fallow" year being used to replenish reserves used up in producing the young; in very cold climates they may only breed every three years. This strictly limits their productivity, and some females may only give birth two or three times during their lifetime. The number of young varies from 3 to 18, and the neonates probably feed on insects, small frogs, and newborn lizards at first.

Numbers of Adders are declining alarmingly in places, with over 70 percent of their numbers lost in the last 20 years in the United Kingdom, for instance, through a combination of habitat destruction, habitat fragmentation, and disturbance. Prejudice also results in the unnecessary killing of Adders. Adder bites are rare and, although they can sometimes be serious, they very rarely result in fatalities.

→ A male Adder, characterized by its boldly contrasting black and gray markings. The markings of females tend to have an overall brownish hue.

Patagonian Lancehead

Small but dangerous

SCIENTIFIC NAME	*Bothrops ammodytoides* Leybold, 1873
FAMILY	Viperidae
SUBFAMILY	Crotalinae
SIZE	16–28 in (40–72 cm), females slightly longer than males
REPRODUCTION	Oviparous
HABITAT	Dry grassland and mountain slopes, coastal sand dunes

The lanceheads, of which there are almost 50 species, are named for their pointed snouts. The larger species are among the most dangerous snakes in Latin America. The Patagonian Lancehead is the smallest species in the genus and is probably the most southerly occurring of all snakes.

This species is endemic to Argentina, where it is wide-ranging, being found in the Andean foothills at elevations of up to 10,000 ft (3,000 m) or more in the west, and at least as far south as the Valdés Peninsula in Patagonia. It is a small, stocky pit viper with a variable pattern. Its dorsal scales are heavily keeled and the scales on the top of its head are small and also keeled.

Much of its range falls within dry grassland or semi-desert habitat, where it lives in gullies amongst rocks, or on rocky mountainsides, but it also occurs on salt flats and sand dunes near the coast. Some of these barren and exposed areas experience extreme diurnal and seasonal temperature fluctuations, sometimes rising to 105 °F (40 °C) in the day and dropping to 15 °F (−10 °C) at night.

Its natural history has been poorly studied but, judging from closely related species, it probably feeds on small rodents, although other prey, such as lizards, may also be taken. It is venomous and irascible; bites are capable of causing fatalities in humans.

The specific name refers to its superficial similarity to some forms of the European Nose-horned Viper, *Vipera ammodytes*, the suffix *–oides* meaning "like" or "similar to."

→ The defensive posture of an alarmed Patagonian Lancehead. This is not a bluff—this species will bite if provoked, and deaths in humans are not unknown.

Desert Horned Viper

Thorny-headed desert dweller

SCIENTIFIC NAME	*Cerastes cerastes* (Linnaeus, 1758)
FAMILY	Viperidae
SUBFAMILY	Viperinae
SIZE	20–24 in (50–60 cm), occasionally longer
REPRODUCTION	Oviparous
HABITAT	Desert

A short, stout viper with a broad, triangular head, perfectly adapted to life in one of the great sandy deserts of the world. Adaptations include sidewinding locomotion (page 59) and producing a hissing sound from the scales on its flanks. Its spotted pale yellow and brown coloration varies according to the color of the sand.

The Desert Horned Viper prefers sparsely vegetated habitats, such as those in the bottoms of wadis (dry riverbeds) where it can find shade in the lee of bushes and grass tussocks. It also shuffles down into loose sand or uses a rodent burrow as a retreat.

The long, thorn-like horns from which it gets its common and scientific names are not present in every individual; unusually some are hornless. There is no obvious explanation for this variation—nor, indeed,

for the horn in those individuals that have them, although a number of theories, such as outline disruption, have been put forward.

Several rows of scales on its flanks have obliquely orientated keels, and each keel is finely serrated. When the snake assumes a C-shaped or horseshoe-shaped coil these oblique keels are at right angles to each other, and as the snake moves its coils they rub against each other to produce a loud hissing or crackling sound. Similar behavior occurs in some carpet vipers, *Echis* species, and in the African egg-eaters, which mimic them. One theory is that producing warning sounds in this way takes the place of normal hissing, during which air, and therefore water vapor, is expelled through the mouth.

All *Cerastes* species, of which there are four in total, lay eggs, an uncommon method of reproduction among members of the Viperidae.

→ At home amongst the sandy wastes of North Africa, the Desert Horned Viper uses its rough scales to produce a loud rasping or hissing sound as a warning.

Black-headed Python

Hidden pits

SCIENTIFIC NAME	*Aspidites melanocephalus* (Krefft, 1864)
FAMILY	Pythonidae
SIZE	5¾–6½ ft (1.75–2 m), exceptionally longer
REPRODUCTION	Oviparous
HABITAT	Dry open forests and grasslands

The Black-headed Python is well named: its head and neck are shiny jet black, contrasting greatly with the rest of its body, which is banded in brown and cream, the bands being brightest in young specimens. It is one of only two pythons that do not have visible heat-sensitive pits in its jaws, the other being the closely related *Aspidites ramsayi* (the Woma). They do, however, have pits inside their mouths and may open their jaws when hunting to bring these into play.

This python spends much of its time underground, in burrows that it takes over from other animals or digs for itself. Basking is initiated by extending its head and neck while keeping the rest of its body hidden within the burrow. As the head warms up, the senses and the nervous system become activated first and the snake is able to react to danger if necessary.

Its diet consists mainly of lizards, including monitors, *Varanus*, large skinks, *Tiliqua*, and bearded dragons, *Pogona*, as well as other snakes. It also takes small mammals and birds on rare occasions, but warm-blooded prey does not figure significantly in its diet, in which case the lack of external heat-sensitive pits is not a problem. Indeed, facial openings of any kind would be a disadvantage for any burrowing reptile as they would soon clog up with soil or sand. Hence there are no burrowing pit vipers, and the only boas that burrow to any extent, the sand boas, also lack pits.

Black-headed Pythons lay proportionately large eggs relative to their body size, and so their clutches are smaller than those of many other pythons of similar size, with most clutches containing fewer than 10 eggs. Based on experiences with these snakes in captivity, females in the late stages of gestation often turn their bodies completely upside down, presumably to expose the developing eggs to as much radiant heat as possible.

→ The impressive Black-headed Python is a terrestrial species from arid parts of Australia, where it hunts predominantly large lizards and other snakes.

Arafura File Snake

Highly adapted water snake

SCIENTIFIC NAME	*Acrochordus arafurae* McDowell, 1979
FAMILY	Acrochordidae
SIZE	Up to 7 ft (2.1 m)
REPRODUCTION	Viviparous
HABITAT	Freshwater lagoons, floodplains, coastal waters

This is an unusual snake with a flattened body, baggy skin, flat head with small eyes on top, and modified scales that are unique to the family. These scales, which give the family its common name of "file snakes," are small and granular. Their ventral scales are reduced to a narrow ridge.

The file snakes have several adaptations to their highly aquatic way of life. Their bodies contain proportionately more blood than other snakes, about 12 percent of their body mass as opposed to 6 percent in typical species. The blood also has a higher proportion of red cells, allowing it to carry more oxygen. They also absorb some oxygen through their skin, which may account for the "bagginess"—as this increases its surface area. The purpose of these adaptations is to allow them to stay underwater for longer periods.

Their scales end in a bristle that is connected to the nervous system and is sensitive to the movement of water, helping them detect the presence of fish, even in muddy conditions. Their rough texture helps them grip the fish on which they feed.

They are long-lived and do not begin to breed until males are five years old and females are seven. They produce large litters of up to 25 young, and females are able to store sperm for at least seven years once they have mated. There may be gaps of several years between litters. Two related species are similar, but the Granulated File Snake, *Acrochordus granulatus*, is boldly banded and lives in coastal waters throughout Southeast Asia and northern Australia, while the other, *A. javanicus*, grows very large and is sometimes known as the Elephant Trunk Snake.

→ The strange-looking, highly aquatic Arafura File Snake is one of only three members of the family Acrochordidae. They have several adaptations to an aquatic way of life, including the position of their eyes and their baggy skin.

Yellow–bellied Sea Snake

An ocean drifter

SCIENTIFIC NAME	*Hydrophis platurus* (Linnaeus, 1766)
FAMILY	Elapidae
SUBFAMILY	Hydrophiinae
SIZE	Up to 3¼ ft (1 m)
REPRODUCTION	Viviparous
HABITAT	Open oceans

A spectacularly patterned snake that is totally adapted for life in the ocean. Unlike other sea snakes, this species drifts on ocean currents, living in the upper layer of the sea, often forming huge aggregations or "slicks." Its nostrils are situated on the top of its snout and can be closed by flaps when it dives, and it can swim forward or backward with equal efficiency.

The body of this sea snake is flattened from side to side and the tail is oar-shaped. Its ventral scales are no wider than those on its back. These adaptations to swimming are so specialized that it is helpless on land if it is accidentally washed ashore.

Its skin attracts algae and barnacles, which it can only rid itself of by shedding, but, because there are no rough surfaces to help in the process, the snake repeatedly loops its body into knots, rubbing one part of its body against another until the skin comes away.

It feeds on pelagic fish, especially those that gather underneath floating debris, which often form large communities, and it ambushes them from the cover of floating vegetation. If it encounters a dense shoal of fish, however, it feeds in a frenzy, striking at any that come within range. Its venom is highly toxic and kills fish quickly and efficiently, while its warning coloration of brightly contrasting black and yellow alerts predatory species that it is venomous and gives it a degree of protection. This species is often taken as by-catch by fishermen and, although they handle them quite roughly, bites to humans are almost unknown.

It is prevented from colonizing the Atlantic Ocean because of the cold currents that move around the southern tips of Africa and South America, although there is a concern that they will eventually find their way through the Panama Canal and into the Caribbean.

This species is often listed as *Pelamis platurus*, the name by which it was known until recently.

→ The extent of the black and yellow markings on the Yellow-bellied Sea Snake is highly variable. Some are almost completely yellow.

Santa Catalina Rattlesnake

A Mexican castaway

SCIENTIFIC NAME	*Crotalus catalinensis* Cliff, 1954
FAMILY	Viperidae
SUBFAMILY	Crotalinae
SIZE	22–29 in (55–73 cm)
REPRODUCTION	Viviparous
HABITAT	Rocky hillsides with thorn and cactus scrub

The Santa Catalina Rattlesnake is endemic to the small island of Santa Catalina in the Gulf of California. It is peculiar among rattlesnakes in having no functional rattle (and is sometimes known as the "rattle-less" rattlesnake). This is thought to be beneficial to the snakes because it enables them to climb up into shrubs and trees in search of roosting birds without inadvertently warning them by rattling.

This is an agile species, more slender than typical rattlesnakes, and therefore better able to climb. They have been found several feet from the ground, gliding quickly along the outer branches of trees; this is very unusual, possibly unique, among rattlesnakes. Their teeth are proportionately longer than those of other rattlesnakes, and this may be an adaptation that enables them to penetrate feathers. They also eat small mammals and lizards. As far as is known, they give birth to quite small litters of young at the end of summer, although records of this rare species are few.

The Santa Catalina Rattlesnake is classified as "Critically Endangered" by the IUCN. The main threats appear to be collection by reptile enthusiasts and commercial collectors, presumably catching them for the pet trade. In the past, predation by introduced cats was a problem, but these have apparently been eliminated. Nevertheless, numbers still seem to be declining.

This species evolved from an isolated population of the Western Diamondback Rattlesnake, *Crotalus atrox*, its closest relative on the mainland, which is present on a small number of neighboring islands.

→ The small island of Santa Catalina, in the Gulf of California, is the only place in the world where the Santa Catalina Rattlesnake can be found. Its many years of isolation have resulted in the reduction, or complete loss, of the rattle at the tip of its tail.

Desert Death Adder

A viper-like elapid

SCIENTIFIC NAME	*Acanthophis pyrrhus* Boulenger, 1898
FAMILY	Elapidae
SUBFAMILY	Hydrophiinae
SIZE	27 in (70 cm)
REPRODUCTION	Viviparous
HABITAT	Desert and desert scrub

Up to nine species of death adders are recognized, found mostly in Australia but also in Indonesia and New Guinea. This is a part of the world that was never reached by true vipers, members of the Viperidae, and so their ecological niche has been occupied by these very atypical elapids.

Everything about the death adders says "viper." They have a short, stout body, broad triangular head, and a short tail. Their fangs are longer and more mobile than those of other elapids, and, like the vipers, death adders are among the few terrestrial elapids to give birth to live young.

Their behavior is also very viper-like, relying on camouflage to ambush their prey. The Desert Death Adder is reddish brown or yellow in color to match the sandy substrate on which it lives and into which it shuffles down if there is no vegetation to hide under. The tip of its tail is often dark and is covered in large, rough scales. When it takes up its ambush position it arranges its body so that its tail is resting near its head. If it detects potential prey, usually desert lizards but sometimes small birds or mammals, it twitches its tail to imitate an insect or insect larva, to entice the victim to within range of its strike. The heavy body provides a firm anchor-point from which to launch its head forward, and it strikes with great speed and accuracy.

Although death adder bites to humans are rare, they can be very serious as their fangs are long and they produce large amounts of neurotoxin, which attacks nerve tissues leading to paralysis and, in extreme cases, death, as the respiratory system fails.

→ The brownish-red coloration of this Desert Death Adder provides it with good camouflage in its desert habitat, while the black tip to its tail can be wriggled enticingly to attract the lizards on which it preys.

REPRODUCTION

Sexual dimorphism

Snakes rarely show sexual dimorphism to the same degree as some other vertebrates, such as many songbirds and lizards, due to the absence of visual displays in their courtship. Anatomically, male snakes have longer tails, thicker at the base, and in some species one sex is significantly bigger than the other.

DIFFERENCES IN COLOR

Difference in coloration between the sexes is restricted to a relatively small number of species. In Boomslangs, *Dispholidus typus*, males in some populations are variable in color and may be bright green, blue, or yellow with black edges to their scales, or uniformly orange or brick-red, whereas females are invariably brown or olive with white or brown undersides.

In Wagler's Pit Vipers, or Temple Vipers, *Tropidolaemus wagleri*, from Southeast Asia, females are black with yellow crossbands around the body, while males and juveniles are green with white spots. In some

Eurasian vipers, *Vipera* species, the dorsal markings of males are darker and there is more contrast between the markings and the background color than in females. This could be because the more muted coloration provides better camouflage, which is especially important for females that need to bask more often when they contain developing embryos.

DIFFERENCES IN SHAPE

Moving on to shape, the strange nasal protuberances found on the Madagascan leaf-nosed snakes belonging to the genus *Langaha* differ between males and females. In *L. madagascariensis*, for instance, the female's nose has a serrated leaf-like structure,

← ↑ Boomslangs, *Dispholidus typus*,
are unusual in being sexually dimorphic.
Females (left) are brown with paler undersides,
whereas males (above) are light green, bright
green or yellow with black edges to their scales,
brick-red, or yellowish-green below and black
above, as here.

whereas that of the male ends in a long tapering proboscis (see page 140). In addition, males are brown with bright yellow undersides and females are light brown with darker mottled markings.

Male boas and pythons have larger spurs than their female counterparts. These are vestigial hind limbs, attached to the pelvic girdle as evidence of snakes' evolutionary history, from the time when their ancestors had limbs. In the species that have them, they have found a secondary purpose: the male uses his spurs to scratch at the cloacal region of the female, stimulating receptive behavior and improving mating success during courtship.

Male garter snakes, *Thamnophis*, have small raised tubercles on their chin and around their cloaca, as does the Turtle-headed Sea Snake, *Emydocephalus annulatus*, and possibly others, that they also appear to use to stimulate females. There is still much to discover about these interesting, but little researched, features.

Mating systems

Most of the snakes that have been studied in detail are seasonal breeders, but this is because they live in parts of the world where researchers are most active. Species in other regions, notably the tropics, breed in response to stimuli that we don't always fully understand.

SEASONAL BREEDING SYSTEMS

Snakes from temperate regions usually mate in the spring, shortly after they have emerged from hibernation. This makes sense, as they often remain near the place where they hibernate at this time of the year, and it is easier for them to find mates before they disperse to the feeding sites. It also gives the females more time to develop eggs or embryos during the warmer months.

Males often emerge first, with females following a few days or weeks later. Females are especially attractive to males immediately after they have shed their skin, which stimulates intense activity amongst nearby males. They will often compete with each other for the opportunity to mate. This may take the form of a free-for-all, known as "scramble competition," or individual combat, which is observed in rattlesnakes and Old World vipers, for example.

In places where the active season is short, females, especially in species that give birth to live young after a long gestation period, may fast for several months while their developing young take up space in their bodies. By the time they have given birth they are very emaciated, and will not be fit enough to breed the following year. These populations will have biennial, or even triennial, breeding cycles, with the years between used to build up reserves.

There are some exceptions to spring matings, where mating takes place in late summer. Some rattlesnakes, hognose snakes, and probably others, may also mate in the autumn, with females retaining sperm and fertilizing their eggs the following year, although they will mate again in spring if the opportunity arises. Sperm storage is a back-up strategy.

← Several males competing for the attentions of a much larger female Green Anaconda, *Eunectes murinus*, form a mating ball.

→ Huge numbers of Red-sided Garter Snakes, *Thamnophis sirtalis parietalis*, emerge from their winter dens simultaneously, with males vastly outnumbering females, leading to a writhing mass of males competing with each other.

SCRAMBLING TO MATE

Garter snakes, *Thamnophis*, of several species engage in scramble competition, in which many males, sometimes in their hundreds, push their way through a writhing mass of rivals for the opportunity to mate with a few receptive females. Once a female has been mated she will leave the area to feed, but other females will emerge from the communal den to take her place, so males are always in a state of high anticipation. When all the females have emerged and mated, both sexes disperse to feed. The females give birth toward the end of summer.

NON-SEASONAL BREEDING SYSTEMS

Seasonal breeding cycles apply particularly to temperate species, because they experience periods of little or no activity during the winter. Hibernation resets their internal clocks each year and breeding is easily synchronized. Snakes from tropical regions, on the other hand, may be active throughout the year, as climatic differences are more subtle.

Relatively little research has been done on the mating systems of tropical snakes, and observations are often made by chance, but some patterns have emerged. Where there are predictable wet and dry seasons this may provide the stimulus for breeding, and mating takes place at the beginning of the wet season, which is usually accompanied by a fall in temperature. Alternatively, mating may take place at the beginning of the dry season so that the eggs or young are produced at the onset of the next lot of rains, when food is often more plentiful. This will vary between species depending on their gestation period and the duration of the wet and dry seasons.

Where there is no distinct wet and dry season, or where the seasons are unpredictable, breeding may be extended throughout most of the year. Many tropical species are probably opportunists. With no seasonal aggregations, mating will take place whenever a male and a receptive female encounter one another.

Reproductive cycles in tropical snakes, then, can be divided into three categories: those that breed throughout the year, those that breed in the wet season, and those that breed in the dry season. Even where a species breeds throughout the year, individual females may only breed once per year (or even once every two years in some larger snakes such as pythons), but, because breeding seasons between individual females are not synchronized, eggs may be produced throughout the year.

→ The Spotted Ground Snake, *Gongylosoma baliodeira*, from Borneo, lives in a region with little seasonal climatic change and probably breeds opportunistically throughout the year.

MALE COMBAT

In many snake species, females are larger than males, sometimes considerably so. This is because they need greater resources to produce eggs or young: the bigger they are, the more eggs they can lay and therefore the more genes they can pass on. Males, on the other hand, which produce sperm, can "afford" to be smaller and therefore to mature at an earlier age.

In species in which males compete physically for the opportunity to mate, however, there will be a selective pressure for them to become larger, the better to defeat their rivals. In this case, males may be larger than females. Male combat has been recorded in many colubrids, boas, pythons, elapids, and vipers, including pit vipers.

Female snakes produce pheromones when they are receptive to mating, and this attracts males. Where more than one male is attracted, combat may ensue. In colubrids, boas, and pythons, males lie parallel to each other and use their heads and necks to attempt to pin their rival's head to the ground. The activity can be repeated several times before a victor, usually the largest snake, is established.

The most dramatic bouts occur among vipers, notably the Old World *Vipera* species and the rattlesnakes, in which the rival males rear up and intertwine their necks and the front halves of their bodies. Each male attempts to force the other to the ground. The bouts may last a few seconds or several minutes, and the combatants may rest between each session. Eventually, one of the males, usually the smaller of the two, accepts defeat and crawls away. The victorious male then returns to the female, and mating may take place.

Where only a small proportion of the females breed each year, males will outnumber females in some years, and competition between males will be intense. In other years, where the number of receptive females is greater, competition is less of an issue and most males may find a mate.

MATING

Mating itself involves the male aligning his body with that of the female until their cloacae (the common opening of the digestive and reproductive tracts) are aligned, before inserting one of his hemipenes (paired sexual organs which are normally inverted and hidden inside the base of the tail) into the female's cloaca. The pair may stay in this position for several minutes during the transfer of sperm. In some species, such as garter snakes, the male leaves a copulatory plug, consisting of a hardened wax-like substance, in the female's cloaca to prevent other males mating with her and displacing the first male's sperm.

← Male Speckled Rattlesnakes, *Crotalus mitchellii*, in combat, with each one trying to press its rival to the ground.

↑ The smaller male Grass Snake, *Natrix natrix*, stimulates the female by rubbing his chin over the top of her head.

Eggs and egg-laying

The period from mating to egg-laying is prone to variation, depending on the temperature and whether fertilization takes place immediately or if the sperm is stored. Most snakes lay their eggs 40–50 days after fertilization. Pythons often take longer than this, sometimes up to twice as long.

EGG SHAPES AND CLUTCH SIZES

The shape of the eggs varies from almost spherical in large snakes to elongated in smaller, thinner species. Small, slender snakes are not able to accommodate eggs with a relatively large diameter, so they lay elongated eggs. In this case, each egg takes up a greater length of the oviduct, and so these species lay smaller clutches. But the relative total weight of these smaller clutches may be the same as that of a large clutch when the size of the female is taken into account; the clutch mass tends to be about 20 percent of body weight across species, regardless of their size, although exceptional values of up to 40 percent relative clutch mass have been recorded.

The largest clutches are those of the largest snakes, the Burmese, African, and Reticulated Pythons (*Python bivittatus*, *P. sebae*, and *Malayopython reticulatus*), all of which have been recorded with clutches of over 100 eggs. At the other end of the scale, many small snakes lay small clutches. The Banded Sand Snake, *Chilomeniscus cinctus*, for example, lays two or three eggs. Female snakes lay increasingly large clutches as they grow, so they may only lay one or two eggs in their first reproductive year but progressively more over their lifetime.

If egg-laying sites are in short supply a number of females may lay their eggs in the same place. Sea kraits, *Laticauda* species, for example, have to come ashore to lay their eggs, and they choose sea caves that are partially above sea level. Freshwater seeping through the roof and sides keeps the humidity high and may also dilute any seawater that enters during storms and tidal surges. The same caves are used every year and may contain large numbers of empty shells from previous clutches.

↖→ The relatively thick-bodied milksnake *Lampropeltis triangulum* produces fairly large clutches of almost spherical eggs (above left), whereas the more slender Leopard Snake, *Zamenis situla*, produces small clutches of elongated eggs (opposite). The hatchlings of both species are similar in size.

← Eggs of the Barred Grass Snake, *Natrix helvetica*, in Devon, UK. This species frequently lays its eggs in dung or compost heaps so the developing eggs can benefit from the heat generated there.

INCUBATION

Snake eggs are enclosed in a parchment-like material that is permeable, allowing the movement of water into the egg from its surroundings. This causes the eggs to expand during incubation, sometimes even doubling in size, or more. To prevent the movement of water out of the eggs, snakes must lay them in moist places which are not prone to drying out. This may be a pile of dead vegetation, damp soil or sand, or a cavity in decaying wood.

Female King Cobras, *Ophiophagus hannah*, build a nest for their eggs by using their coils to gather a heap of dead leaves. Pythons and some other species remain

↑ A small python, *Antaresia childreni*, coiled around her clutch. By doing this, she can disguise the bright white eggs and guard them from potential predators, a rare example of maternal care in snakes.

→ A Leopard Snake, *Zamenis situla*, emerges from its egg, using a tiny egg-tooth on its upper jaw to slit the shell. It will have been developing for about two months.

with their eggs during incubation, but most female snakes, having found a suitable site, lay their eggs and then take no further interest in them. Where they occur, alligator nests, consisting of piles of rotting reeds and other vegetation, are used by snakes, as are ant nests and termite mounds in some parts of the world. These are thought to make especially good incubation chambers, as the insects maintain them at a constant temperature, and keep them free of mold.

EGG BROODING

Certain pythons are unusual among snakes in brooding their eggs. The female maneuvers the eggs into a heap as she lays them. Once the clutch is complete, she coils herself around them. In this position, she is able to protect them from predators, and some species can also produce metabolic heat to speed up the incubation. Brooding females twitch and shiver at intervals, depending on the ambient temperature. The cooler it is, the more often they shiver. In this way they can maintain a constant temperature of 88–91 °F (31–33 °C) within the clutch. Brooding behavior also helps to prevent heat loss to the atmosphere, and females may bask for short periods before returning to the eggs and transferring heat to them. Thermal control has been established for several species, including the Burmese and Indian Pythons, *Python bivittatus* and *P. molurus*, and may occur in other, as yet less well-studied, species.

Live young

Viviparous species, those that give birth to live young, are widely distributed amongst the colubrids, the boas, and the vipers. All the acrochordids and the sea snakes (but not the sea kraits) are also live-bearers. Because these groups are unrelated, viviparity is assumed to have evolved several times in snakes.

Giving birth to live young comes with huge advantages to snakes in certain groups. Species that live in cold climates can bask and move about to optimize their body temperature to the benefit of their developing young. Significantly, the snakes that occur furthest north and south in the world, and at the highest altitudes, are all live-bearing vipers (except the Hot Springs Snake, *Thermophis baileyi*, which is a special case because it lives in a warm microclimate). The reproductive strategies of even closely related species may differ if they live in different climatic zones.

Finding a suitable egg-laying site can be problematic for species that live in certain habitats, such as very dry places and very wet places, so they will tend to be live-bearers. Many desert species, including the vipers, sand boas, and Australasian death adders, *Acanthophis*, which are actually elapids, are live-bearers, as are the fully aquatic marine snakes and the freshwater-inhabiting file snakes, *Acrochordus*. The North American water snakes, *Nerodia*, and the garter snakes, *Thamnophis*, from the same region, are all live-bearers.

Live-bearing snakes may retain the embryos in their oviducts until they are fully formed, nourishing them via a placenta, in which case they are said to be viviparous, or the embryos may subsist on a yolk within an egg sac which they break out of just before or just after parturition, in which case the species is said to be ovo-viviparous.

LITTER SIZES

Litter sizes in viviparous species are similar to clutch sizes of egg-layers, with the largest-known broods being those of the Puff Adder, *Bitis arietans*, and the Lancehead, *Bothrops atrox*, both purportedly capable of producing 100 or more young. Among colubrids, there are records of 95 and 101 young respectively for the African Mole Snake, *Pseudaspis cana*, and the Common Garter Snake, *Thamnophis sirtalis*.

← A Copperhead, *Agkistrodon contortrix*, emerging, still enclosed in its egg-sac, while another neonate, born previously, struggles to free itself.

LIVE YOUNG OR EGGS?

Amongst the smooth snakes of Europe
and North Africa, the northern species,
Coronella austriaca, is viviparous,
whereas the Southern Smooth Snake,
C. girondica, is oviparous, reflecting the
different environments in which they live.

→ Two members of the same genus,
Coronella, have different breeding strategies:
the Smooth Snake, *C. austriaca*, from northern
Europe (inset), gives birth to fully formed young
whereas the Southern Smooth Snake,
C. girondica, from warmer regions, lays eggs.

Parthenogenesis

Parthenogenesis is the ability to reproduce without mating. It is not uncommon in invertebrates, and there are also a number of parthenogenetic lizards, but, among snakes, there is just one species that reproduces exclusively this way.

The Brahminy Worm Snake, *Indotyphlops braminus*, is thought to have originated in India, but it has been introduced to many other places scattered around the tropical and subtropical world, often as a stowaway in the soil of potted plants—hence its alternative name of "Flowerpot Snake." It is a female-only species, and individuals, or eggs, transported accidentally to other parts of the world, can go on to produce fertile eggs that will, in turn, hatch into more females. The process of parthenogenesis has therefore facilitated its spread, because only a single individual, or a clutch of eggs, needs to be present.

← As far as is known, the Brahminy Blind Snake, or Flowerpot Snake, *Indotyphlops braminus*, is the only parthenogenetic snake.

↗ This Yellow-bellied Water Snake, *Nerodia erythrogaster flavigaster*, appears to have given birth without needing to find and mate with a male. This sometimes occurs where males are not available (in captivity, for instance) and is known as facultative parthenogenesis.

Although parthenogenesis as a strategy may be beneficial to a species' survival and distribution in the short term, it limits natural variation because all the genetic input is from a single individual. Each of the offspring, all female, will be clones of their mother (and of their grandmother, and so on) and there will be little or no variation between individuals. Variation, however, is the driving force behind evolution and adaptation. As conditions change, populations with a wide range of characteristics are better able to adapt, because at least some of them will be suited to the new conditions. Parthenogenetic populations, of identical individuals, may be less able to adapt.

An alternative type of parthenogenesis has also been recorded in snakes more recently. This is facultative parthenogenesis, in which females that would normally breed by mating may, in the absence of males, produce eggs or young. Facultative

parthenogenesis has long been suspected in snakes, ever since techniques for breeding them in captivity were developed in the 1970s. Usually, females that laid eggs or gave birth in the absence of a male partner were thought to have stored sperm from previous matings, perhaps before they were captured. This explanation, however, is not consistent with snakes that were born or hatched in captivity and reared in the absence of a male, but which nevertheless went on to produce offspring of their own, of either sex. Facultative parthenogenesis is the subject of ongoing research.

Growth and development

Snakes have indeterminate growth; that is, they continue to grow throughout their lives, albeit slowly once they reach reproductive size. This may occur in less than one year in some small tropical species, but it can take more than five years in the case of large species, and in snakes from temperate regions, which may be inactive for half the year.

Male snakes tend to mature earlier than females, although sometimes they mature at roughly the same size; there are no cases of females maturing at a younger age than males. Even in species in which males eventually reach a larger size than females, such as most rattlesnakes, males still reach maturity soonest but then carry on growing. By contrast, once females start to reproduce, their growth slows down or ceases altogether. In evolutionary terms, it pays females to wait until they have reached a larger size before reproducing because then they will be able to produce more—or larger—young or eggs.

Captive snakes potentially reach breeding size earlier than their wild counterparts, because they have

abundant food, fewer parasites, and a longer active season. Even large pythons, for example, can reach breeding size in less than two years, a fraction of the age at which their wild counterparts would begin breeding.

←← A hatchling Grass Snake, *Natrix helvetica*, crawls out of its egg-shell.

← A number of hatchling Grass Snakes, *Natrix helvetica*, begin to disperse during what will be the most vulnerable time of their lives. Only one or two, if any, are likely to make it to adulthood.

↑ A Variable Bush Viper, *Atheris squamiger*, going through the process of shedding its skin.

SHEDDING

As snakes grow they must shed their skin, and the frequency with which this takes place depends on their growth rate. The first shed usually occurs within a week or so of birth or hatching. After this shed, the skin is less permeable to water, protecting the young snake from dehydration.

Feeding rarely takes place until after the first shed. If a young snake is successful in finding food, it may shed two or three more times before it enters hibernation, assuming it is a temperate species. Snakes that hatch late in the season, such as the Barred Grass Snake, *Natrix helvetica*, may not feed at all until they emerge from hibernation the following year.

COLOR CHANGE WITH AGE

Juvenile snakes are sometimes marked differently to adults. The reasons for this are not always clear, but juvenile snakes are obviously more vulnerable to predation than adults and so they may benefit from more effective camouflage, especially if their habits and habitats change as they grow.

In some of the ratsnakes, *Elaphe* and *Pantherophis* for instance, the young are blotched at hatching but the blotches gradually fade as they begin to mature over their first year or two, to be replaced by plain coloration or longitudinal stripes. The change takes place in stages, each time the snake sheds its skin. In the related Ladder Snake, *Zamenis scalaris*, from western Europe, the juveniles have a series of bold H-shaped markings down their backs, but as they grow the central bar of each H fades, leaving two longitudinal lines.

The juveniles may mimic venomous vipers from the region (and they certainly behave like vipers), but once the adults approach maturity they are too large to be convincing viper mimics.

Overall darkening of the colors occurs in a number of species, but most notably in some forms of milksnakes, *Lampropeltis triangulum, L. polyzona,*

and others, in which the hatchlings are almost unbelievably brightly colored but gradually darken as they grow until, in some forms, they become almost uniformly black.

The most dramatic color changes are probably those of the Emerald Tree Boa, *Corallus caninus,* and the Green Tree Python, *Morelia viridis.* In a remarkable example of convergence (see page 52), both these green snakes start life as bright yellow juveniles (rarely red or brown in the case of the python) but change gradually to bright green during the first year or two of their life. No convincing explanation has been given for this phenomenon. Another arboreal boa, *Sanzinia madagascariensis,* is dark green or olive as an adult but newborn young are reddish-brown.

↖← As an adult (above), the Yellow Ratsnake, *Pantherophis quadrivittatus,* is yellow with four faint longitudinal stripes along its body, but juveniles (below) are boldly blotched. The markings begin to change after their first year, and the change is complete after 2–3 years, depending on their growth rate.

↑ Green Tree Pythons, *Morelia viridis,* are bright yellow (occasionally red) when they hatch but have changed to green by the time they are two years old.

Madagascan Leaf-nosed Snake

Sexually dimorphic vine snakes

SCIENTIFIC NAME	*Langaha madagascariensis* Bonaterre, 1790
FAMILY	Pseudoxyrhophiidae
SIZE	Up to 3¼ ft (1 m)
REPRODUCTION	Oviparous
HABITAT	Tropical dry and wet forest

This species, and two that are closely related to it, are remarkable for several reasons, most notably of which is their sexual dimorphism. Males are brown with yellow undersides, with a white stripe running down the flank where the two basic colors meet, whereas the females are mottled gray or brown. Uniquely, they have nasal appendages that also vary between the sexes, that of the male being long, pointed and spear-shaped whereas the female's is flattened from side to side and has a series of serrations along the top.

The purpose of these appendages is completely unknown, although they clearly enhance the snake's crypsis. Two poorly known related species, *Langaha alluaudi* and *L. pseudoalluaudi*, also have nasal appendages and show sexual dimorphism; the females have, in addition to their nasal appendages, small leaf-like "horns" above their eyes. All *Langaha* are extremely long and slender, living in low vegetation, usually at about 5–6 ft (1–2 m) above the ground. They remain motionless, which makes them difficult to see, as their coloration matches that of the vines and shrubs in which they live.

Langaha madagascariensis is widely distributed in Madagascar and is found in lowland forest throughout much of the island, although the species is rarely seen and is poorly known. They have vertical pupils, which would indicate a nocturnal lifestyle, although some observers have noted that they have diurnal habits. A small number of gravid females have been collected and have laid clutches of 5–12 eggs. Adults and young feed on prey they catch in bushes, presumably consisting mostly of small lizards such as geckos and, possibly, frogs. They are rear-fanged snakes, of no danger to humans.

→   The female Madagascan Leaf-nosed Snake is one of the more bizarre snakes from Madagascar. The purpose of its strange snout may simply be to disguise its outline when resting in trees and shrubs.

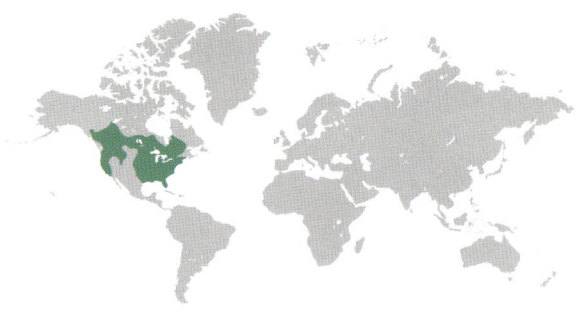

Common Garter Snake

Versatile and variable

SCIENTIFIC NAME	*Thamnophis sirtalis* (Linnaeus, 1758)
FAMILY	Colubridae
SUBFAMILY	Natricinae
SIZE	Females to 3 ft (90 cm), males smaller
REPRODUCTION	Viviparous
HABITAT	Damp or wet places, often near streams and ponds

The Common Garter Snake has a huge range over North America, with up to 12 subspecies recognized. These range from being common and widespread to rare and of limited distribution, and vary in color and markings. There are few parts of North America that are not home to one form or another. They are absent only from the central, drier parts of the continent.

The most studied subspecies is the Red-sided Garter Snake, *Thamnophis sirtalis parietalis*, the most northerly occurring subspecies and famous for its spring mating aggregations in Manitoba, Canada, in which hundreds of males, freshly emerged from hibernation dens, compete for the attentions of the larger females, crawling over each other in a frantic competition to pass on their genes. This subspecies has dull red bars along either side of a pale vertebral stripe, a theme that is also present in several other subspecies.

Some of the western subspecies, notably the San Francisco Garter Snake, *T. s. tetrataenia*, have more extensive and brighter red markings on their sides, and red heads. This subspecies has a very limited range and is protected by law as it is threatened by habitat loss from agriculture and urban development, as well as by collection for the pet trade.

Behavior and seasonal activity of the species depend very much on where it lives; in the north of the range it hibernates for up to six months of the year and may emerge when snow is still on the ground, whereas in the south, such as in Florida, it may not hibernate at all. Similarly, subspecies from cold climates may only breed every two or three years, whereas those from warmer regions breed every year. Truly a very versatile and varied snake.

→ The Common Garter Snake exists in many forms, of which this one, the very rare San Francisco Garter Snake, *Thamnophis sirtalis tetrataenia*, is arguably the most colorful.

Western Diamond-back Rattlesnake

Largest rattlesnake in southwest America

SCIENTIFIC NAME	*Crotalus atrox* Baird & Girard, 1853
FAMILY	Viperidae
SUBFAMILY	Crotalinae
SIZE	5 ft (1.5 m) but occasionally to more than 6½ ft (2 m)
REPRODUCTION	Viviparous
HABITAT	Dry places such as deserts, grasslands, coniferous woodlands

This is a large and impressive rattlesnake, quite common in the American Southwest, and probably responsible for the majority of human snakebite deaths in North America. Conversely, humans are this species' most important predators, and huge numbers are killed by cars, habitat destruction, and persecution by narrow-minded people, especially through rattlesnake "roundups" in some states.

This is a wide-ranging and fairly common species in drier habitats, normally seen in rocky, sparsely vegetated places such as hillsides and gullies. In the spring it is active during the day and may be discovered basking on paths and roadsides. Later in the summer it becomes gradually more nocturnal, retreating to rock crevices or animal burrows during the heat of the day. In autumn it becomes day-active again, and this period often coincides with rainy weather and an increasing abundance of rodents, at a time when the young rattlesnakes are born. Adults eat a huge variety of prey, including prairie dogs, rabbits, and ground squirrels, as well as ground-nesting birds, lizards, and amphibians.

During the winter it seeks out large underground hibernacula, or dens, and 100 or more may congregate at the same site, sometimes in the company of other species of snakes. This, of course, makes them vulnerable to so-called "hunters," who may use gasoline or exhaust gases to force them out of the dens.

When alarmed, a Western Diamondback Rattlesnake will raise its head and up to one-third of its body off the ground, draw its head back and face the threat while raising its tail and rattling loudly and continuously. If the warning is not heeded (and it usually is!) the snake will strike. It says something for the tolerance of the species, however, that bites are rare, with less than a handful of fatalities recorded each year.

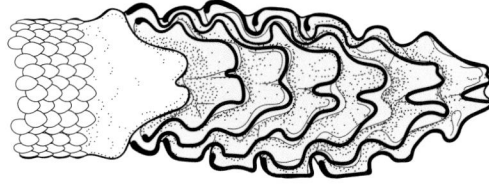

Rattle cross section
Each segment is shaped to grip the one in front of it.

→ Western Diamondback Rattlesnakes are formidable reptiles when they assume a defensive posture, like this one, with its head raised and the rattle "sounding off" loudly.

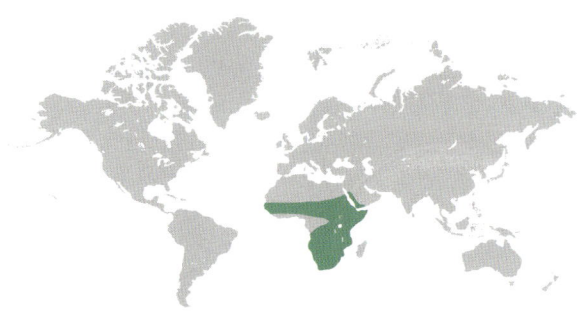

Puff Adder

A snake best avoided

SCIENTIFIC NAME	*Bitis arietans* (Merrem, 1820)
FAMILY	Viperidae
SUBFAMILY	Viperinae
SIZE	2½–3 ft (80–90 cm), occasionally to 5 ft (150 cm)
REPRODUCTION	Viviparous
HABITAT	Varied, including dry grasslands, scrub, and rocky hillsides

Probably the most feared snake in the areas where wit occurs, which are extensive. It relies on its cryptic coloration and is easily overlooked, often resulting in contact with humans. When threatened, it inflates its body with air and exhales in a loud, and alarming, drawn-out hiss or a series of puffs. It will strike rapidly if this warning is ignored, often with serious consequences.

Although it is primarily nocturnal, the Puff Adder will often bask in the open during cooler weather in the south of its range. It rarely climbs, and typically rests with its body coiled at the base of a shrub, from where it ambushes its prey. Its heavy body provides a solid anchor from which to throw its head forward, and it strikes hard and fast. Small prey may be held and swallowed as soon as they stop struggling, which usually happens very quickly, but large prey is released and tracked down later. Puff Adders can eat relatively large prey items owing to their wide, spade-shaped heads and thick bodies, and they will take mammals, birds, other reptiles, and amphibians. There is a record of one eating a small tortoise.

This is a very prolific snake, with litters averaging 20–40 young but exceptionally over 100. The largest litter, of 156 young, born in a zoo, is the largest recorded for any snake.

Bites from Puff Adders are fairly common. A full bite contains enough venom to kill 1–3 humans and results in local pain, swelling, and bruising. Because the venom is slow-acting, however, medical attention is normally available before irreversible damage is done. Death only occurs in 5–10 percent of cases.

→ The Puff Adder is a familiar sight in parts of southern Africa, where it is responsible for a number of medically important bites each year. Its habit of lying concealed and motionless contributes to the likelihood of it being stepped on.

BOA CONSTRICTOR

Boa Constrictor

The stuff of legends

SCIENTIFIC NAME	*Boa constrictor* Linnaeus, 1758
FAMILY	Boidae
SUBFAMILY	Boinae
SIZE	10 ft (3 m), rarely to 13 ft (4 m) or slightly more. Often exaggerated!
REPRODUCTION	Viviparous
HABITAT	Tropical rainforests, deciduous forests, savanna, secondary forests, and human-altered habitats such as plantations

Perhaps the best-known snake in the world, at least in popular literature, and almost unique in having identical common and scientific names. This is the legendary constrictor from South America with which many explorers have embellished their stories of derring-do over the centuries.

The Boa Constrictor, also known as the Common Boa or Red-tailed Boa, is a semi-arboreal snake of varied habitat, normally found in forests and wooded savanna but adapting to altered environments and often to be found around towns and villages. It feeds almost exclusively on mammals up to the size of opossums and monkeys, and birds, but is one of the few boinine boas to lack visible heat pits. When hunting, it may use an ambush technique or actively seek its prey.

These snakes are powerful constrictors, striking rapidly, grasping their prey with long, backwardly pointing, curved teeth and immediately throwing several coils of their body around it. It may take several minutes for death to occur through asphyxiation. After a large meal, Boa Constrictors retire to a hollow tree, cave, or other hidden place while they digest it, which can take up to a week depending on its size.

Although they are mostly active at night, I have seen one successfully stalk and catch one of a large flock of parrots at a clay lick on an Amazonian riverbank in the middle of the day and then climb into a tree to bask. This snake was apparently a regular visitor to the site.

The populations occurring further north, into the drier habitats of Central America as far north as Mexico, are now considered to be a separate species, *Boa imperator*.

→ The Boa Constrictor is the archetypal snake of the tropics, very versatile and with a wide range of habitat preferences, including villages and even cities.

Green Anaconda

The jungle giant

SCIENTIFIC NAME	*Eunectes murinus* (Linnaeus, 1758)
FAMILY	Boidae
SUBFAMILY	Boinae
SIZE	10–13 ft (3–4 m), exceptionally to 16 ft (5 m); females much larger than males
REPRODUCTION	Viviparous
HABITAT	Semi-aquatic, living in swamps, marshes, lagoons, and slow-moving streams and rivers

Probably the largest snake in the world, the Green Anaconda is the subject of countless tales of exaggeration, some of which are so fantastical that they hardly deserve consideration. Others may be more accurate, but difficulties in measuring huge snakes—and in transporting the evidence, living or dead, from such remote and steamy places—make it hard to arrive at a universally accepted measurement.

The largest size for a Green Anaconda is generally acknowledged to be 17 ft 1 in (5.21 m). This snake weighed 214 lb 15 oz (97.5 kg). Males are much smaller than females, rarely attaining 10 ft (3 m) and a fraction of their weight. The Green Anaconda is exceeded in length only by the Reticulated Python (page 66), which is, however, much more slender. Adult Green Anacondas eat a variety of prey, from fish to large mammals such as tapirs,

Tapirus, and Jaguars, *Panthera onca*. They share their habitat with caiman (members of the alligator family), which they also eat.

Males often form large mating "balls," with up to a dozen homing in on a receptive female and attempting to mate, usually in shallow water. Each male attempts to use his cloacal spurs to stimulate the female and effect a mating, a contest that may continue for several days or even weeks.

The gestation period is 6–7 months, and litters of up to 100 have been recorded, although 20–40 is more usual. A recent instance of facultative parthenogenesis (see page 135) has been recorded for this species. A cryptic species, *Eunectes akayima*, has been described recently (2024), from the northern part of the species' range, based on genetic profiling. It is indistinguishable from *E. murinus* in appearance.

→ More at home in the water than on land, the Green Anaconda is the world's heaviest snake. It shares its habitat with crocodilians and semi-aquatic mammals, both of which form part of its diet.

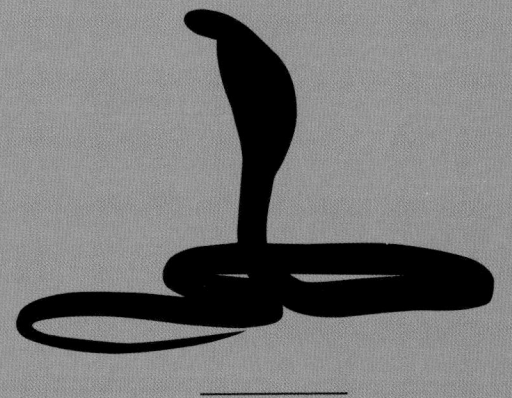

DIET AND
FEEDING

A carnivorous lifestyle

All snakes are carnivores, feeding on a variety of other animals. Collectively they will feed on almost anything, dead or alive, that will fit into their bodies, ranging from worms, mollusks, and insects to hoofed mammals. Some, however, have more specialized requirements than others.

The feeding behaviors of snakes differ from those of other animals in many ways, and are the result of several unique factors. Their elongated, limbless body shape makes capturing, overpowering, and swallowing prey a highly specialized process. First, the snake has to find its prey, using a combination of sight, scent, and, sometimes, heat detection. Then it must overpower it without sustaining an injury. And finally, it has to swallow an item that may be several times the diameter of its own head and body.

Constrictors grip their prey and instantly coil around it, gradually tightening their coils until it can no longer draw breathe. Others have evolved venom of various types with which to overcome their prey, and these species may also have modified teeth, to deliver the venom. Some venomous snakes strike and then hold their prey until it is dead, but others release the prey and track it down after it has succumbed, using their various sensory systems to do so.

← The arboreal and nocturnal Black-headed Cat Snake, *Boiga nigriceps*, swallowing a Harlequin Flying Tree Frog, *Rhacophorus pardalis*, that it has snatched from a leaf or branch.

↗ This very large African Rock Python, *Python sebae*, has overpowered a small antelope and will proceed to swallow it, an operation that can take several hours.

Overleaf: Southern Smooth Snake, *Coronella girondica*, overpowering and swallowing a Slow-worm, *Anguis fragilis*.

JAWS THAT ARE NOT JOINED

Snakes' jaws are very loosely articulated, allowing them to open up to an astonishing amount. This is achieved by moving the two halves of the lower jaw away from each other and also from the tooth-bearing upper jawbones, so that they can be spread widely to accommodate large prey. During swallowing, the snake "walks" its mouth over the prey, using teeth to pull it in one side at a time. After swallowing, snakes often maneuver these bones back into position by opening their mouths and flexing their jaws.

Swallowing may take many minutes or even hours. Digesting large prey can take anything from a few days to a week or more, during which time the snake itself may be vulnerable to attack, as its mobility is impaired.

Snakes belonging to the oldest families, especially the Leptotyphlopidae (thread snakes) but also, to a lesser extent, the blind snakes, shield-tails, and sunbeam snakes, have more limited movement of the jaws, and these species eat smaller prey.

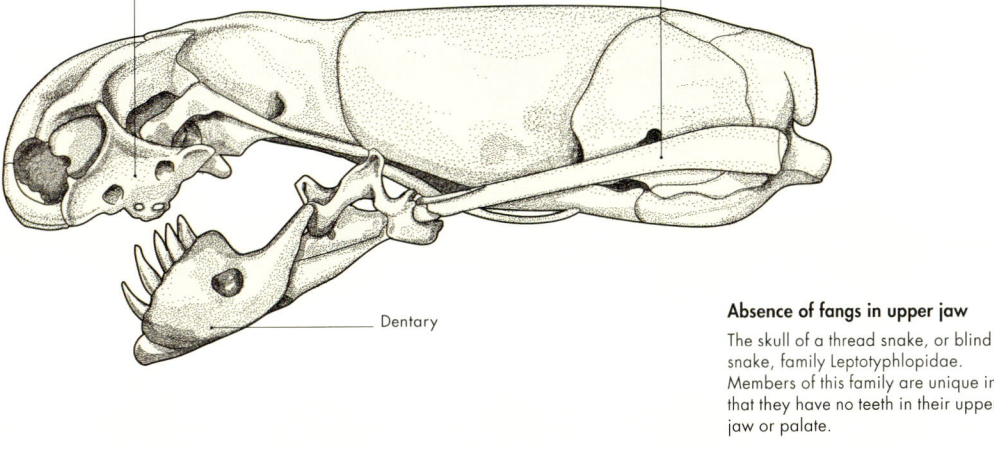

Maxilla

Quadrate

Dentary

Absence of fangs in upper jaw
The skull of a thread snake, or blind snake, family Leptotyphlopidae. Members of this family are unique in that they have no teeth in their upper jaw or palate.

GENERALISTS AND SPECIALISTS

Most species of snakes will eat a variety of prey, the only requirement being that they can find and overpower it. These species' diets are more dependent on what prey is available than on preference, and their "preferred" prey may simply be a species that happens to be abundant where they live.

For example, the abundance of *Anolis* lizards on Caribbean islands makes them a convenient and plentiful prey species, and many snakes eat them, typically stalking them at night when the lizards are sleeping on leaves or stems. These snakes could be considered lizard specialists, even though they may opportunistically take other prey if it presents itself.

Some species, though, have staked their survival on the continued availability of a particular type of prey. At some point along the evolutionary pathway these snakes have taken a turn that leads them to become specialists rather than generalists; they have become so locked into a certain type of prey that other options are no longer open to them. They may have modified behavior and dentition, and search in specific places where their "chosen" prey is found. The prey in which they specialize must be abundant enough to sustain them, and this is often only possible under certain circumstances, which is why there are only a relatively small number of specialists among the 4,000 species of snakes.

↖ An African egg-eater, *Dasypeltis* species, has engulfed a very large egg, and will proceed to saw through the shell, releasing the contents into its gullet.

↗ An Asian Blunt-headed Slug Snake, *Aplopeltura boa*, eating a snail by levering the fleshy part of the body out of the shell.

→ A Beaked Sea Snake, *Hydrophis schistosus*, searches among rocks for its prey, consisting of small fish, especially eels.

Examples include the 18 species of African egg–eating snakes, *Dasypeltis*, that will only eat birds' eggs. Their dentition and skeleton are highly modified to this end, and they are unable to eat any other prey. Some rainforest snakes only eat land mollusks (slugs and snails) and, again, they have modified skulls. Certain sea snakes belonging to the genus *Hydrophis* have extremely narrow heads and necks so that they can poke about in burrows in search of the eels on which they exclusively feed. All these snakes, and others like them, have evolved in such a way that suits them well for their chosen prey but makes it difficult or impossible to switch to another type.

Invertebrate prey

Many small snakes eat soft-bodied invertebrates of various kinds, from ants to earthworms to snails. Some larger snakes focus on crustaceans, and they have evolved a range of adaptations that allow them to specialize on their preferred prey species.

The approximately 470 species belonging to the Scolecophidia have small mouths and rigid jaws, so most of them are unable to eat anything larger than the ants and termites among which they dwell, although a few of the larger species will also take beetle larvae and small spiders.

Earthworm specialists include the shield-tailed snakes, Uropeltidae, from Sri Lanka and southern India, and the *Toxicocalamus* species from New Guinea, many of which appear to eat nothing else; many other snakes include earthworms in their diets. Slugs and snails are the preferred food of a number of species. The North American Sharp-tailed Snake, *Contia tenuis*, appears to eat only slugs and their eggs; other slug-eaters include Kirtland's Snake, *Clonophis kirtlandii*, and the Red-bellied Snake, *Storeria occipitomaculata*, which eat slugs and earthworms. There must be many more across the world that eat prey of this type, but the diets of most small snakes are poorly known.

Snails are more difficult for snakes to tackle, as they have no means of crushing their shells. Even so, two groups of snakes, one from tropical America, including members of the genera *Dipsas* and *Sibon*, and another from South Asia, in the genera *Aplopeltura*, *Asthenodipsas*, and *Pareas*, are snail specialists and have modified jaws adept at extricating the soft body from its shell (see page 178). The African *Duberria* species eat mostly slugs but may also take snails on occasion, although they appear to have no modifications for winkling them out of their shells.

Insects and arachnids are widely eaten, by some specialists as well as by generalists. There are 11 African snakes known as the centipede-eaters, *Aparallactus* species; as their name suggests, they feed

SPECIALIZED INVERTEBRATE EATERS

Some species of *Leptotyphlops* are so specialized that they produce a substance in their cloacal glands that contains pheromones that suppress the aggressive behavior of soldier termites, while the Texas Blind Snake, *Rena dulcis*, shown here, does not swallow whole ant or termite larvae but squeezes them in its mouth, forcing out the soft internal parts, which are swallowed while the exoskeleton is discarded.

exclusively on centipedes, which they subdue with a venomous bite. In North America, the hook-nosed snakes, *Gyalopion* and *Ficimia*, numbering nine species altogether, eat spiders, centipedes, and scorpions.

Finally, the Asian mudflat-dwelling Crab-eating Snake, *Fordonia leucobalia*, eats crabs, which it dismembers by gripping them in its coils and pulling them apart with its mouth, while in North America Graham's Water Snake, *Regina grahami*, eats crayfish, but only those that have recently molted.

← A Spotted Coffee Snake, *Ninia maculata*, from Central America, a small, terrestrial snake, typical of the many species that feed on soft-bodied invertebrates.

↗ *Geophis sartorii* is known as the Terrestrial Snail Sucker on account of its specialized diet.

→ The Queen Snake, *Regina septemvittata*, is a highly specialized feeder on crustaceans, and its diet consists almost entirely of newly molted crayfish.

Vertebrate prey

Generally speaking, vertebrates are larger than invertebrates (although there are exceptions). Snakes that eat vertebrates therefore tend to be the larger species, and subduing, handling, and swallowing them becomes more of a challenge, requiring a range of different techniques.

FISH

Specialist fish-eaters are found in the Homalopsidae, Acrochordidae, and Elapidae. Homalopsid water snakes include the unusual Tentacled Snake, *Erpeton tentaculatum*, which uses the sensory tentacles on its snout to locate nearby fish (page 188). Species such as *Cerberus*, *Enhydris*, and *Homalophis* live in shallow freshwater lakes and ponds, including fish-farms, or in estuaries, and have

enlarged fangs in the rear of their mouths to envenomate and subdue their prey. The three species of file or wart snakes, *Acrochordus*, are totally aquatic; two live in rivers, freshwater lakes, or brackish estuaries, and the third, *A. granulatus*, is marine. All eat fish, and they are covered in rough, tubercular scales ending in stiff bristles, which help them grip their slippery prey and hold it until they locate it with their mouth.

Many of the sea snakes are crevice feeders, poking their heads into holes in rocks and coral, or into sand burrows, in their search for resting fish, especially eels and gobies. A number of species have narrow heads and long thin necks, probably adapted for this activity. When they find prey, whether in a crevice or in open water, they strike and hold onto it until their venom takes effect, before turning it around so they can swallow it head-first; many fish have sharp dorsal spines that would damage the snakes if they tried to swallow

← A Red-sided Garter Snake, *Thamnophis sirtalis parietalis*, eating a fish.

↑ This European Dice Snake, *Natrix tessellata*, has caught a large freshwater fish and will proceed to swallow it. Only the snake's ability to stretch its jaws widely will make this a realistic proposition.

them backward. Three sea snakes, *Aipysurus eydouxii*, *Emydocephalus annulatus*, and *E. ijimae*, have even more specialized diets, feeding exclusively on fish eggs, which they search for among crevices and burrows.

Snakes with heat-sensitive pits are not adapted to feed on cold-blooded prey such as fish, but the American Cottonmouth, *Agkistrodon piscivorus*, is an exception, as its specific name suggests (see page 182). This snake eats a variety of food items including fish. In the Florida Keys some populations feed heavily on marine fish spilled by adult seabirds returning with food for their young.

Many of the aquatic and semi-aquatic colubrids, especially the water snakes, garter snakes, keelbacks, and others, which form the subfamily Natricinae, are fish-eaters, although some of them also eat other prey such as amphibians and earthworms.

AMPHIBIANS

Frogs, toads, and salamanders make up the diet of a great many snakes, most of them generalists, but there are also a few specialists. Snakes in several genera are attracted to large breeding aggregations of frogs and toads. In tropical regions, where some arboreal frogs lay their eggs on leaves and branches overhanging pools, these may also fall prey to snakes.

Amphibians are most common in moist habitats, where snakes hunt them on the ground, below the ground, in water, or in trees and bushes. Snakes such as *Nerodia* and *Natrix*, some garter snakes, *Thamnophis*, and many others, are semi-aquatic, and frogs and toads figure highly in their diet.

TOXIC TOADS …

Many toads and salamanders secrete toxic substances from glands in their skin, giving them a certain amount of protection against predators. Bufonid toads have large poison glands behind their heads and, while some snakes avoid them, others, such as the European grass snakes, *Natrix*, will take them on occasion, and are seemingly immune to the toxins. The tropical frogs belonging to the Dendrobatidae (poison dart frogs) and Mantellidae (mantellas), and some others, advertise their toxic nature by bright coloration; they are so poisonous that all predators avoid them completely. Similarly, the black and yellow European Fire Salamanders, *Salamandra salamandra*, are not troubled by snakes.

← A young Rhombic Night Adder, *Causus rhombeatus*, attempting to feed on a toad, its preferred prey, although this one looks as though it might be just too big for it!

↑ A Grass Snake, *Natrix natrix*, hunting a green frog.

... AND EASIER PICKINGS

Other than these species, amphibians are relatively defenseless and make easy prey in places where they are plentiful. They are simply grabbed and swallowed in most cases. The African night adders, *Causus* species, hognose snakes of the North American genus *Heterodon*, and the South American Hognose Snake, *Xenodon dorbignyi*, are toad specialists and have upturned rostral scales that they use to root out toads buried in the ground. The *Heterodon* species, and some of the others mentioned, have enlarged, backward-pointing fangs at the rear of their mouths, mounted on mobile bones that can be swung downward, bringing the fangs into play; it is thought that these fangs are used to puncture the skin of toads, which often inflate themselves to prevent predators from swallowing them.

Eating reptiles

Reptiles make up a large proportion of snakes' diets, perhaps the largest. Small lizards such as geckos and skinks, amphisbaenians, and smaller snakes are plentiful in many places, and they are the ideal shape to fit into snakes' elongated bodies.

↙ Galapagos Racers, *Pseudalsophis biserialis*, feast on newly hatched Marine Iguanas, *Amblyrhynchus cristatus*, during the breeding season.

→ King Cobras, *Ophiophagus hannah*, are specialist snake-eaters. This one is swallowing a venomous pit viper.

Amphisbaenians (sometimes known, inaccurately, as worm lizards) are all burrowing reptiles. Some of the coral snakes, *Micrurus* species, specialize in eating them, as do the African *Chilorhinophis gerardi*, and the quill-snouted snakes, *Xenocalamus*. All these slender snakes are well suited to following amphisbaenians along their burrows and consuming them underground.

Lizards occur everywhere that snakes do. In places such as Australia and the Caribbean, lizards are by far the most abundant vertebrate and therefore the most common food; some snakes will eat nothing else. The seasonal availability of hatchling lizards is a feeding bonanza for some snakes. On Fernandina in the Galapagos Islands, the Western Racers, *Pseudalsophis occidentalis*, pursue hatchling Marine Iguanas, *Amblyrhynchus cristatus*, over the rocks. In Central America, the Mexican Burrowing Python, *Loxocemus bicolor*, eats lizard eggs and ambushes hatchling Green Iguanas, *Iguana iguana*, (and marine turtles) as they emerge from their nests.

Many of the New World coral snakes, *Micrurus*, eat other snakes, some exclusively so. The King Cobra,

Ophiophagus hannah, is a confirmed snake-eater, as its generic name suggests. Related species including the kraits, *Bungarus* species, and the Asian coral snakes, *Maticora* and *Calliophis*, are also snake eaters. In North America the small Short-tailed Snake, *Stilosoma extenuatum*, from Florida, appears to eat only snakes, predominantly the black-headed snake *Tantilla relicta*. Small snakes are swallowed live, but larger individuals are constricted.

Reptile eggs are eaten by numerous species; specialists include the Scarlet Snake, *Cemophora coccinea*, the Asian kukri snakes, *Oligodon*, the desert banded snakes, *Simoselaps*, in Australia, and the shovel-snouted snakes, *Prosymna*, in Africa. All of these snakes have upturned snouts, thought to help in bringing the sharp, rear cutting teeth into play without first having to engulf the egg, which can be relatively large. Once the shell is slit, the contents of the eggs are swallowed.

Very few snakes are large enough to take adult crocodilians, but they are a favored prey of Green Anacondas, *Eunectes murinus*, and large pythons in Africa and Asia may take them if the chance arises.

Bird-catchers

Birds are not easy prey for snakes. Catching them is an obvious problem, and their thick covering of feathers makes it hard for a snake to get a good grip on them. Bird-eating snakes, then, tend to be specialists, although a few species eat them opportunistically.

Arboreal snakes, such as the tree boas, *Corallus* species, and the Green Tree Python, *Morelia viridis*, often have very long teeth, enabling them to penetrate plumage. They hunt by hanging head-down from a branch with several bends in their neck and the front part of their body, ready to be straightened out quickly as they strike. From this position they can catch birds in flight or birds alighting on a lower branch. They use their heat-sensitive facial pits to detect the presence of birds nearby.

Other snakes hunt birds when they are in the nest and unable to escape. In Europe, the Montpellier Snake,

Malpolon monspessulanus, and the Ladder Snake, Zamenis scalaris, have been found in the burrows of European Bee-eaters, *Merops apiaster*, after gorging themselves on chicks, and European vipers, *Vipera* and *Macrovipera*, will eat nestling birds in the spring: the habitat of these species has few trees, so most birds nest on the ground or in bushes where they are accessible to the snakes.

In the Bass Strait, between the Australian mainland and Tasmania, the Tiger Snakes, *Notechis scutatus serventyi*, on Mount Chappell Island feed heavily on chicks of the Short-tailed Shearwaters (or "Muttonbirds"), *Ardenna tenuirostris*, that nest on the island in high densities. Adults eat so many birds that they can store enough energy to last them until the next breeding season. Juveniles, which are too small to take the chicks, feed on lizards, switching their diet when they are large enough.

The Brown Tree Snake, *Boiga irregularis*, is a generalist that feeds on birds if the opportunity presents itself. This has led to disastrous consequences on the island of Guam, where it was introduced accidentally; this is described in more detail on page 268.

← Chappell Island Tiger Snakes, *Notechis scutatus serventyi*, gorge themselves on nestling Muttonbirds during the breeding season.

→ This vine snake, *Ahaetulla* species, has caught a small bird and will proceed to swallow it. Only its excellent camouflage makes stalking alert prey such as this a possibility.

Mammals

Only the larger snakes can eat adult mammals. Their size and shape makes them difficult to swallow, and they are capable of fighting back and causing injuries unless an efficient method of subduing them is available.

The vast majority of snakes that prey on mammals are either constrictors or venomous. Small snakes that swallow their food live will occasionally take young rodents if they come across nests, but this method of feeding is ill-suited for eating larger mammals that are potentially armed with claws and teeth. On the other hand, many places have an abundance of mammals, especially rodents, rabbits, and hares, and they are a good source of protein for snakes that are large enough and equipped to handle them.

The various genera that go under the name of ratsnakes in North America, Europe, and Asia, as well as a host of other medium-sized constrictors, feed on small to medium-sized mammals and may consume a large proportion—up to 40 percent in some cases—of the population of rodents such as mice, rats, rabbits,

↖ A small python from Australia, *Antaresia childreni*, has constricted a small mammal and is pulling its body from its coils as it swallows it.

← Brown House Snakes, *Boaedon capensis*, perform a useful service for the farming community in southern Africa by preying on mice and rats in barns and grain stores.

↗ An African Python, *Python sebae*, swallowing a small gazelle which it has constricted. Swallowing and digestion will take a considerable time.

and ground squirrels every year. The very large constricting snakes, such as boas and pythons, can tackle and swallow animals up to the size of deer, pigs, and antelopes, although large meals adversely affect their ability to move about and may make them prone to predation themselves. There are also cases where large snakes have died as a result of the horns of prey animals piercing their bodies from the inside.

Other mammal-eaters subdue their prey by injecting it with venom. These snakes may hold the prey until the victim dies, or they may strike and release it immediately, going in search of it a few minutes later, when the venom will have taken effect. Which method is used will often depend on the size of the prey. Vipers, with their stout bodies and broad heads, can manage proportionately larger prey than most other snakes. The pit vipers, including rattlesnakes, have evolved heat-detecting organs that are adaptations to hunting warm-blooded prey, as have some boas and pythons.

Hunting strategies

Hunting methods used by snakes fall into two main categories: active hunters that go in search of their prey, and ambushers that wait for the prey to come to them, sometimes known as "sit-and-wait" predators. These are two extremes, however, and there are intermediate strategies that might suit a snake's particular situation or choice of prey.

ACTIVE HUNTERS

Diurnal active hunters tend to be slender, fast-moving snakes with large eyes and good vision. American and European whipsnakes and racers, African sand snakes, boomslangs, and mambas, and the Australian elapids also known as whipsnakes, are all good examples of this type. They eat a variety of prey, but lizards are predominant, probably because they are also diurnal and because the snakes' slender bodies are better suited to swallowing them.

Nocturnal active hunters prowl on the ground in search of terrestrial mollusks, frogs, and other small animals, and in bushes in search of sleeping lizards. They use stealth to approach their prey, and the arboreal species are particularly long and slender with deep bodies that enable them to stretch out between branches. They typically have large eyes and vertically elliptical pupils, which can be closed down more effectively than round pupils, thus protecting them from strong light during the day.

← Well camouflaged, heavy-bodied, and with a small head that can be launched forward rapidly, the Blood Python, *Python brongersmai*, is well equipped to ambush its prey.

↗ A Péringuey's Adder, *Bitis peringueyi*, lurks beneath the surface of the Namib Desert, with only its eyes showing, ready to ambush any small lizard that comes within range.

SIT-AND-WAIT PREDATORS

Ambush predators tend to be short and thick-bodied. Their heavy bodies give them a good anchor-point from which to throw their head forward, and they can strike in the blink of an eye. They are able to overpower and swallow relatively large prey; this is important, because ambushing is unpredictable and they have to seize any opportunity that comes their way to grab a meal. They are invariably well camouflaged in order to go undetected. Vipers, such as the Gaboon Viper, *Bitis gabonica*, and other *Bitis* species, the rattlesnakes and lanceheads, *Bothrops*

species, and the Malayan Pit Viper, *Calloselasma rhodostoma*, are the archetypal sit-and-wait predators; their lives consist of many hours coiled in one place waiting for suitable prey to come along. When it does, they strike rapidly and may release it immediately, preferring to track it down after it has died and ceased to be a threat. In Australasia, where there are no true vipers, the death adders, *Acanthophis*, which are actually elapids, behave in exactly the same way.

← A Pope's Pit Viper, *Trimeresurus popeiorum*, dangling and twitching its red tail in the hope of attracting a small bird, lizard, or frog.

→ The tail of the remarkable Spider-tailed Viper, *Pseudocerastes urarachnoides*, is adorned with long, filament-like scales that the snake moves in a life-like fashion to imitate a scuttling spider. Its prey consists largely of small migratory birds looking for an easy meal.

↘ When hunting, death adders, *Acanthophis* species, position their grub-like tail near their head, in readiness to strike at any small animal that approaches.

LURING

Some ambush predators improve their chances of catching prey by encouraging it to come within striking range. They achieve this by using the tip of the tail, which is often of a different color to the rest of the body, and moving in an enticing, worm- or caterpillar-like manner.

This technique is used mostly by vipers. Typically, these species are cryptically colored apart from the tail, and they hide among dead leaves or stones so that their body is not obvious. They position their tail so that it is close to their mouth, within easy striking distance, and may raise it off the surface before wriggling it about. Luring is well documented in New World species such as the Cantil, *Agkistrodon*

bilineatus, juvenile Copperheads, *A. contortrix*, and small rattlesnakes, *Sistrurus*, among many others. The Northern Death Adder, *Acanthophis praelongus*, a viper-like elapid from Australia, uses its tail as a lure in the same way.

Of the Old World vipers, the most remarkable example is the Spider-tailed Viper, *Pseudocerastes urarachnoides*, described as recently as 2006 from western Iran. The tip of its tail is modified into a knob-like structure and the area immediately in front of this is covered in scales that are drawn out into filaments, the whole structure resembling a large spider when twitched that attracts small birds.

Whether or not luring is effective depends on the intended prey. Lizards, frogs, and birds are the animals most likely to investigate a small wriggling object, mammals less so. Juvenile snakes are much more likely to use caudal luring than adults, perhaps because they have different prey species or because their smaller tails better resemble the worms and grubs that are the models for the lures. In these species the tail may be brightly colored in the juveniles but gradually change to match the rest of the body in adults.

Snake venom

Venomous snakes occur in several families. The venom they produce is primarily used to kill or incapacitate their prey but may also be used in defense. The amount of venom, and its composition, varies between families and species.

↓ A European Asp Viper, *Vipera aspis francisciredi*, from northern Italy. Bites from this subspecies, though serious, are milder than those from other populations, such as that in the Pyrenees, which can produce life-threatening bites.

→ A Common or Spectacled Cobra, *Naja naja*, a widespread Asian species that can give a fatal bite.

The most important venomous snakes, in terms of numbers of species and the sophistication of their venom-delivery apparatus, belong to the families Elapidae (the cobra family) and Viperidae (vipers). The snakes in these families, totaling nearly 800 species in all, have fangs situated at the front of their mouths where they can be easily deployed. Another group of snakes in the Colubridae, including the Boomslang, *Dispholidus typus*, and many others, have large grooved fangs at the rear of their mouths that can introduce venom into a wound by chewing. The 70 species of so-called mole vipers, usually placed in the Atractaspidae but sometimes regarded as part of the Lamprophiidae, are also venomous and have, between them, a variety of fang types with which to inject their prey. Many species, however, are small and feed on invertebrates.

The venom itself consists of a cocktail of substances, each of which produces a different effect on the victim. In very broad terms, these can be divided into neurotoxic enzymes and proteins, which affect the nervous system and which are typical of elapid snakes, and proteases, enzymes that damage the walls of blood vessels, preventing coagulation. The latter are also known as hemostatic venoms, and they are typical of vipers. This is very much a generalization, however. Snakes from both families often produce venoms with several types of enzymes and proteins present. Snake venoms are highly complex; their composition can vary between species and, sometimes, between populations and even individuals.

The amount of venom a snake injects depends on several factors such as the size and health of the snake and the type of prey, which can differ from place to place. It is in the snake's interests to use enough venom to kill the prey without "wasting" it by injecting more than is necessary. This topic, and the evolution of immunity in some prey species, is the subject of current research.

Catesby's Snail-eating Snake

Snails are on the menu

SCIENTIFIC NAME	*Dipsas catesbyi* (Sentzen, 1796)
FAMILY	Colubridae
SUBFAMILY	Dipsadinae
SIZE	2½–3 ft (80–90 cm)
REPRODUCTION	Oviparous
HABITAT	Primary and secondary rainforest, plantations

Probably the most common and widespread of the South American snail-eaters, this species is slender and arboreal, with a blunt head, feeding mostly on tree snails, for which it is highly adapted.

Eating snails is a specialized business. Snakes are not able to swallow them whole or digest their shells, so the dedicated snail-eaters have evolved a method of extracting them from their shells. First, the snake coils its body around the snail's shell so that it is held firmly. Next, it pushes its lower jaw, which is narrow and has long curved teeth, into the shell while the snout remains in contact with the outside of the shell to provide purchase. Once the lower jaw is fully inserted, the teeth hook into the snail's fleshy body and gradually pull it out, in much the same way as humans may use a bent pin to extract the flesh of whelks or winkles.

The process is aided by secretions from specialized glands under the tongue, which appear to paralyze the snail and make extraction easier. The whole operation takes only about two minutes. In addition to snails, Catesby's Snail-eater will also eat slugs and other soft-bodied invertebrates such as worms, and amphibian eggs, although snails make up the majority of its diet.

Convergent snakes from Asia, belonging to the family Pareatidae, have similarly blunt snouts and narrow mandibles and, in addition, an asymmetrical arrangement of teeth on their lower jaw which favors snails whose shells have either right- or left-handed twists (dextral or sinistral). The bias in the snakes' jaws will depend on the most frequently encountered form of snail, which varies from place to place.

→ Tree snails are the preferred prey of the highly arboreal Catesby's Snail-eating Snake, although it may also eat slugs.

CHIONACTIS OCCIPITALIS

Western Shovel-nosed Snake

A sand swimmer

SCIENTIFIC NAME	*Chionactis occipitalis* (Hallowell, 1854)
FAMILY	Colubridae
SUBFAMILY	Colubrinae
SIZE	12–18 in (30–45 cm)
REPRODUCTION	Oviparous
HABITAT	Sandy and gravel deserts

This small snake is adapted to a subterranean lifestyle, having smooth, shiny scales, and an obvious underslung lower jaw. Seen in profile, its snout tapers sharply from a point roughly above its eyes to form a sharp edge, resulting in the "shovel-nosed" characteristic that gives it its name. This shape enables it to "swim" through loose sand, usually just below the surface. Its nostrils are equipped with valves that can close them off to prevent sand particles from entering while it is burrowing.

The Western Shovel-nose lives in the American Southwest, in parts of the Mojave and Sonoran Deserts wherever there is suitable sandy habitat. Dry river washes with sparse vegetation are a favorite haunt. It is largely nocturnal, and seems to specialize in hunting arachnids (spiders, scorpions, and

solifuges) and centipedes but will also take other invertebrates including ant pupae, beetle larvae and adults, and cockroaches. Most prey items are simply grasped and swallowed, although they may be moved around in the jaws until better aligned.

Captive individuals were seen to lie on or just under the surface of sand in the evening and lift their heads and necks when potential prey was nearby. If the prey strays within range they will strike from this position, or they may actively chase the prey. They appear to be immune to the venom of scorpions.

This is a colorful snake. Its markings consist of about 30–35 black bands on a white or cream background and, in most forms, there are narrow red or orange bands in the spaces between them. In some populations, however, especially those of the Mojave Desert, the red bands are faint or absent. The bold markings may serve to mimic the venomous coral snakes, although its range does not overlap any of these species so it is more likely that the pattern performs a startle function when the snake moves rapidly. It breeds in the spring, laying 2–4 elongated eggs which hatch in about 60 days.

Wing-like processes prevent excessive lateral movement

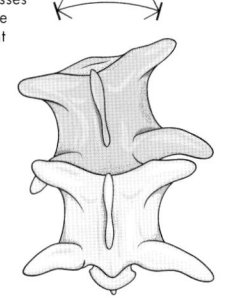

Interlocking snake vertebrae
The loosely interlocking processes on a snake's vertebrae prevent the backbone from twisting too far and damaging the spinal cord.

→ A Western Shovel-nosed Snake in the Colorado Desert, where it is a common but rarely seen inhabitant of areas where sand and fine gravel accumulate, such as dry river beds known as "washes."

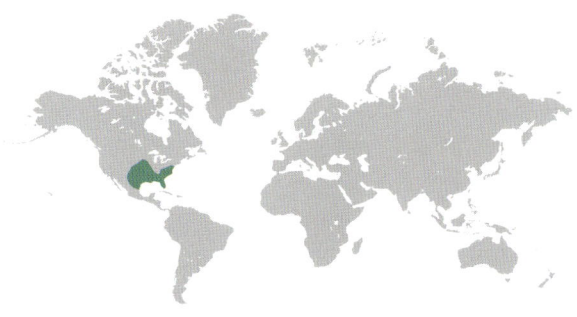

Cottonmouth

The snake that eats anything

SCIENTIFIC NAME	*Agkistrodon piscivorus* (Lacépède, 1789)
FAMILY	Viperidae
SUBFAMILY	Crotalinae
SIZE	2¼–3 ft (65–90 cm), occasionally much longer
REPRODUCTION	Viviparous
HABITAT	Semi-aquatic in swamps, coastal marshes, ponds, and ditches

A large, heavy-bodied venomous snake with heavily keeled scales, which, when threatened, opens its mouth widely, displaying the pale interior from which it gets its name. It can be aggressive and dangerous, and human fatalities have occurred. Due to its excellent camouflage and apparent lethargy, it can be easily overlooked and trodden on.

This species changes from banded tan and black as a juvenile, to almost totally black by the time it is adult. It basks at the edges of swamps and riverbanks, often remaining immobile until the last minute. Upon discovery, or if it is found out in the open, it flattens its body and forms a tight coil, striking out repeatedly.

Cottonmouths, also know as Water Moccasins, are very common in suitable places, no doubt due in part to their adaptability. They are diurnal during the spring and autumn but switch to a nocturnal activity pattern in the warmer months. They are more tolerant of cold weather than many other snakes, and hibernate only during the very coldest months.

Cottonmouths may swallow their food as soon as they capture it, or they may release it after envenomating it and track it down after it has died, depending on its size and its ability to fight back. The Cottonmouth eats an astonishing variety of food. This includes many species of fish—hence the specific name *piscivorus*—but also salamanders, frogs, juvenile alligators, turtles, lizards, snakes, birds and their eggs, and many different mammals. Some of the fish that it eats are marine species, and it has been seen scavenging fish spilled from the nests of seabirds, even eating partly decomposed, dried-out fish carcasses. It also forages in the water underneath heron, ibis, and stork rookeries, on the lookout for young birds that have fallen out of the nests.

→ Cottonmouths can be hard to spot when they are resting among vegetation, but they open their mouths widely to display the white interior if they are disturbed.

Blunt-headed Tree Snake

The bootlace snake

SCIENTIFIC NAME	*Imantodes cenchoa* (Linnaeus, 1758)
FAMILY	Colubridae
SUBFAMILY	Dipsadinae
SIZE	4 ft (1.25 m)
REPRODUCTION	Oviparous
HABITAT	Lowland rainforest and plantations

Probably the world's most slender snake, known locally as the "Fiddlestring Snake," this species has a huge range from Central America to northeastern Argentina, wherever there is suitable habitat. It is largely arboreal, although it sometimes also forages on the forest floor. A row of very large hexagonal dorsal scales running along the ridge of its back is a distinctive feature.

Like many arboreal snakes, this species has a body that is flattened from side to side (laterally compressed), which gives it added rigidity when it is reaching out across spaces between branches. This also allows it to snatch diurnal lizards, especially anoles and related species, that sleep on leaves. It has large eyes with vertically elliptical pupils, closing down to a narrow slit in daylight. They can be directed downward,

presumably improving its field of vision when hunting amongst branches. During the day it coils up in the centers of bromeliad plants, in other epiphytes, in tree holes, or under forest litter.

Apart from lizards it also feeds on small frogs, such as the numerous *Eleutherodactylus* species (coquis) that live in the same part or the world, and tree frogs, Hylidae. Small items of prey are seized and swallowed live, but larger items may be constricted.

Females reach breeding size in about two years, and breeding takes place throughout the year. In keeping with its slender shape, *Imantodes* lays elongated eggs, about three times as long as their diameter, in clutches numbering from one to eight, but most commonly two or three. Hatchlings measure about 12–14 in (30–35 cm) in length.

→ This Blunt-headed Tree Snake is just about as thin as a snake can be, accentuating its wide head and large eyes.

Northern Cat-eyed Snake

An adaptable night-hunter

SCIENTIFIC NAME	*Leptodeira septentrionalis* (Kennicott, 1859)
FAMILY	Colubridae
SUBFAMILY	Dipsadinae
SIZE	2–3 ft (60–90 cm)
REPRODUCTION	Oviparous
HABITAT	Tropical and subtropical forest

A very common nocturnal species, mostly terrestrial but climbing into low shrubs and bushes in search of food. It is a nocturnal prowler and eats a wide variety of frogs, toads, lizards, and small snakes. Salamanders, small fish, and mice are taken occasionally.

This species is best known for its depredations on the eggs of leaf-nesting tree frogs belonging to the Phyllomedusinae, including the Red-eyed Frog, *Agalychnis callidryas*, and its relatives, and the Hourglass Treefrog, *Dendropsophus ebraccatus*. These frogs lay their eggs on leaves and branches overhanging ponds, allowing them to develop out of water and away from aquatic predators. The tadpoles drop into the water below as they hatch. When the frogs are breeding, which they do explosively, the snakes will move about in the bushes in search of mating pairs and freshly laid egg-masses. Huge amounts of spawn are taken, often 50 percent or more of all eggs laid, and individual snakes may return to the bushes several times to plunder the same egg-masses.

They lay 6–10 eggs in leaf-litter or tree holes. A related species, *Leptodeira annulata*, is known to lay its eggs in the large nests of leafcutter ants, and *L. septentrionalis* may do likewise. The young measure about 8 in (22 cm) in length and are more brightly marked than the adults.

This is a rear-fanged snake that chews its prey to introduce its venom, which is fairly powerful and can cause a painful local reaction in humans, but it is not considered dangerous.

→ The Northern Cat-eyed Snake is frequently seen prowling around the edges of small ponds, especially those with overhanging vegetation, where leaf frogs may be laying eggs.

Tentacled Snake

One of a kind

SCIENTIFIC NAME	*Erpeton tentaculatum* Lacépède, 1800
FAMILY	Homalopsidae
SIZE	2 ft (60 cm), occasionally longer
REPRODUCTION	Viviparous
HABITAT	Totally aquatic, in ditches, slow-moving streams, and the like

The only member of its genus, the Tentacled Snake is unique in possessing a pair of protrusions (the "tentacles") on the end of its snout. These contain many nerve endings which enable it to detect small changes in water pressure, such as those caused by a fish swimming nearby.

In cross-section this species is almost rectangular, being flattened above and below, and it holds its body rigidly if picked up. Its pattern may consist of longitudinal stripes or blotches in shades of brown, tans, and cream, arranged along its body. Its scales are heavily keeled, and the ventral scales are reduced to a narrow row. It lives in murky, stagnant bodies of water, such as lakes, paddy fields, and fish-farms, and never voluntarily leaves the water. It is almost helpless on land.

Its most distinctive features, the "tentacles," have been the subject of much speculation in the past, but recent research indicates that they are highly sensitive organs, used in a unique method of hunting. It rests with its neck in a J-shaped position so that its head is pointing back along its body. When a fish swims between its head and body it first startles it and then anticipates its escape route, striking into that space. It is very effective; fish are often caught and swallowed live in a single strike.

Although it is apparently common, commercial fishing in some of the places where it lives probably impacts its population, because the snake is taken as by-catch, and is eaten in small numbers by humans.

→ The unique and bizarre-looking Tentacled Snake is a confirmed fish-eater with a flattened body and reduced ventral scales. Its pattern and markings provide it with effective camouflage in the weedy waters in which it lives.

Southern Hognose Snake

A comical bluffer

SCIENTIFIC NAME	*Heterodon simus* (Linnaeus, 1766)
FAMILY	Colubridae
SUBFAMILY	Dipsadinae
SIZE	20 in (50 cm), occasionally longer
REPRODUCTION	Oviparous
HABITAT	Dry places with sandy soil, including cultivated fields and pine and oak woodlands

This and the other hognose snakes, of which there are four species in total, all found in North America, are easily recognized by virtue of the spade-like upturned rostral scale at the tip of their snout that gives them their common name.

Hognose snakes use their upturned snout like a plow to force their way into hard ground when hunting for their preferred prey of toads, including spadefoot toads, *Scaphiopus*, which burrow down into the soil during dry weather to avoid desiccation. They have specially adapted fangs in the rear of their mouth, and a gap between these and the rest of their teeth, so that their specialized long rear fangs can be swung down when needed. They use these fangs to puncture the toads, which often inflate their bodies when they are

threatened. The fangs also deliver a mild venom, produced in their Duvernoy's glands. Bites to humans are uncommon but can result in some pain, with accompanying swelling and bleeding.

The hognose snakes are nature's clowns. If threatened they will flatten their head to make themselves look like a viper, puff their body up, hiss, and make mock strikes with their mouth closed. The dorsal markings give them a superficial resemblance to small rattlesnakes from the same region, and their theatrical displays are surely an attempt to mimic them. If all else fails, they may feign death by flipping their body upside down and allowing their tongue to loll from their mouth, sometimes accompanied by forcing blood from their mouth. Once the threat has passed they slowly "come back to life" and right themselves.

→ The Southern Hognose Snake is the smallest of the hognose snakes. It feeds largely on toads, using its upturned snout to dig down into the soil where it buries itself during dry weather.

DASYPELTIS SCABRA

Common Egg-eating Snake

An egg specialist

SCIENTIFIC NAME	*Dasypeltis scabra* (Linnaeus, 1758)
FAMILY	Colubridae
SUBFAMILY	Colubrinae
SIZE	Up to about 3¼ ft (1 m), usually much smaller
REPRODUCTION	Oviparous
HABITAT	Varied, from dry stony plains and hillsides to forests

Although a number of snakes eat birds' eggs as part of their diet, the 18 species of African egg-eating snakes and a single species from India, *Elachistodon westermanni*, are the only snakes that feed exclusively on them. The Common, or Rhombic, Egg-eater is by far the most familiar species. It is a slender snake with a rounded snout and rough scales.

The Common Egg-eater is highly adapted to its diet. Its teeth are reduced in number and in size, so they don't impede the movement of the eggs into the snake's mouth. The jaws can be enormously distended to engulf the egg, which is usually pushed against the side of the nest to give purchase. Once in the mouth, the egg is passed into the throat, which is equipped with a row of elongated, downwardly pointing processes (the hypapophyses) from the lower surface of the vertebrae, which are used to saw through the shell of the egg. The snake works its throat backward and forward until the egg collapses. The contents are swallowed and the characteristically folded empty shell is regurgitated.

Because their prey is seasonal, egg-eaters gorge themselves during the spring and summer. Small birds, such as weaver birds, Ploceidae, are abundant where they live and occur in large colonies, making it relatively easy to move from nest to nest.

The Common Egg-eater is a convincing mimic of the venomous snakes that live in the same region, such as the Common or Rhombic Night Adder, *Causus rhombeatus*, or the Horned Adder, *Bitis caudalis*. Like several vipers, they can form their body into a horseshoe-shaped coil and rub their serrated scales together to create a rasping noise (see page 106). They may also open their mouth widely to display the black interior.

→ The wide-ranging Common Egg-eater has evolved a number of adaptations to equip it for eating birds' eggs, on which it feeds exclusively. None is more remarkable than the extent to which it can stretch its jaws to swallow them.

Brown House Snake

An effective pest controller

SCIENTIFIC NAME	*Boaedon capensis* Duméril, Bibron & Duméril, 1854
FAMILY	Lamprophiidae
SUBFAMILY	Lamprophiinae
SIZE	2–2½ ft (60–80 cm), occasionally to 3¼ ft (1 m)
REPRODUCTION	Oviparous
HABITAT	Grasslands, including semi-arid and rural habitats

As their name suggests, house snakes often live in proximity with humans, in houses and outbuildings, where they benefit from the concentrations of rodents that also find these places attractive. For this reason they do a useful job in controlling vermin and are often tolerated by the human inhabitants. They are also found in open country, in fields, grassland, and semi-desert scrub.

This snake is slender, muscular, and more-or-less cylindrical in cross-section, with a narrow head and smooth, satin-textured scales. It is variable in color, and may be mid- to reddish brown, becoming darker with age. The ventral scales are white or cream, and strongly iridescent.

These are nocturnal, terrestrial prowlers, rarely climbing. They are powerful constrictors and can handle relatively large prey, proportionately as large as many boas and pythons. Their generic name means "boa-teeth" and is very apt: their teeth are long and strongly recurved for holding on to struggling prey. After striking, they rapidly throw many coils around their victim. Struggling prey is quickly subdued and the snake begins to swallow it head-first, easing the body from its coils as it does so. Although rodents are by far the most common prey, they will also eat other small mammals (including bats), as well as birds and lizards.

Breeding takes place in the spring and summer, and females may lay several clutches in a single season; in captivity they have laid up to six clutches. There are up to 18 eggs in a clutch, but on average about eight, and they take 2–3 months to hatch. The young snakes are small and slender but easily capable of swallowing nestling rodents, and they can consume a whole litter in one visit.

→ The Brown House Snake is a common species in farmlands and urban surroundings, as well as more natural habitats, in southern and parts of eastern Africa.

LATICAUDA COLUBRINA

Yellow-lipped Sea Krait

A banded aquatic elapid

SCIENTIFIC NAME	*Laticauda colubrina* (Schneider, 1799)
FAMILY	Elapidae
SUBFAMILY	Laticaudinae
SIZE	3–6½ ft (1–2 m), females much larger than males
REPRODUCTION	Oviparous
HABITAT	Coastal waters and reefs

The Yellow-lipped Sea Krait is an amphibious species, living on rocky and coral outcrops around coastlines or islands. Compared to the true sea snakes, the eight species of sea kraits are not as highly adapted to marine life because they need to come ashore to lay eggs and shed their skin. Like them, however, they do have a paddle-shaped tail for swimming.

The behavior of this species is something of a mystery, owing to the difficulty of studying them, but they appear to congregate in certain places at certain times, perhaps for breeding. The factors controlling the timing of these events are not clear, although they are most likely to come ashore during the night, especially if this coincides with high tide.

Courtship takes place at the high tide line, where males intercept the larger females coming ashore and attempt to mate with them; several males may be entwined around a single female. Females lay 7–13 eggs in sea caves or crevices above the high water limit. Freshwater from rain and seepages maintains a high humidity and prevents them from drying out.

This species eats mostly eels, actively hunting for them among reefs by poking its small head into holes and crevices. Once it has fed it comes ashore to bask, drink freshwater, and shelter in hollow logs and holes in rocks. Its venom is very potent and, although they are reluctant to bite, a small number of incidents have been recorded, mainly involving fishermen clearing them from their nets. They should be regarded as dangerously venomous.

→ An inhabitant of coastal waters and coral reefs, the Yellow-lipped Sea Krait is one of the more widespread of the sea kraits, occurring over large parts of Southeast Asia.

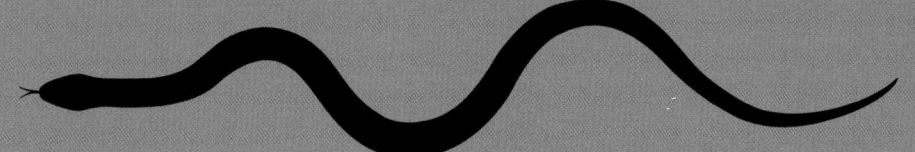

ENEMIES AND
DEFENSE

Enemies

Snakes have many predators, not least of which are other species of snakes. Other animals that prey on snakes may do so incidentally, or they may specialize in snakes, or in all reptiles.

Among these are birds of prey such as the Secretary Bird, *Sagittarius serpentarius*, from Africa, serpent eagles, *Circaetus* species, from Africa and Europe, several owls, and the Road Runner, *Geococcyx californianus*. Large corvids (crows and ravens) will take snakes opportunistically, as will the widespread Crested Caracara, *Caracara plancus*, and many other species of eagles and hawks.

Mammals such as foxes, racoons, wolves, and even shrews and rodents will also take snakes; hibernating snakes and hatchlings are especially vulnerable. Mongooses of various species are traditional enemies of snakes in Asia and Africa but have been introduced to other parts of the world where they have wrought havoc with the native snake populations, which have no defense against them.

← A Laughing Kookaburra, *Dacelo novaeguineae*, with a small snake in its bill. Uncountable small snakes fall prey to birds, some of which specialize in hunting them.

↗ A Slender Mongoose, *Galerella sanguinea*, attacking a Boomslang, *Dispholidus typus*, in southern Africa. The snake has inflated its throat to make it appear bigger than it is.

→ Even large amphibians, like this African Bullfrog, *Pyxicephalus* species, will eat small snakes, including hatchlings and newborn young of some of the larger species.

Large lizards such as the monitor lizards, *Varanus*, in the Old World, and tegus, *Tupinambis*, in the New World, frequently eat snakes. The worm and thread snakes are, as their name suggests, small and worm-like in appearance and, where they occur, they will be potential prey of anything that eats this type of food; this includes frogs and toads, and even invertebrates such as spiders and centipedes. The young of larger species may also fall prey to predators in this category. There is a record of a group of 16 small Rinkhals cobras, *Hemachatus haemachatus*, presumably newborn litter-mates, taken from the stomach of a preserved African Bullfrog, *Pyxicephalus adsperus*.

With such a wide array of potential predators, it is not surprising that snakes have evolved many strategies for defending themselves, some of them unique.

Defense strategies

Most snakes, sensing danger, try to hide away. Their shape lends itself to sliding between or beneath rocks or logs, or into crevices. Almost uniquely in the animal kingdom, an individual snake can take many forms, from outstretched like a piece of rope, to a tightly coiled lump, and every shape in between. This makes it difficult for would-be predators to build up a search image.

ESCAPE

Active snakes can often outpace predators, and once they have been discovered they take off rapidly, often heading for a retreat such as a crevice or burrow they have used before. Snakes are especially adept at squeezing into small spaces.

Slender, streamlined snakes are clearly much faster than heavy-bodied species, and are therefore more inclined to "run" if disturbed. Arboreal species can likewise move rapidly through their three-dimensional habitat. Having said that, the speeds at which snakes move is often exaggerated; this is partly because of the optical illusion created by their bodies as they undulate rapidly across the ground, seeming to mesmerize the observer until suddenly they are gone. This applies especially to snakes that have longitudinal markings along their bodies.

In fact, snakes are not very speedy compared with animals that have legs. Even so, a snake moving at a speed of, say, 7–8 mph (12–13 km/h), which appears to be the maximum they can attain, will still outpace a human or some other animals, especially over broken ground such as scrub or scree. Indeed, the rougher the ground, the faster the snake and the slower the pursuer.

As a last resort, a few snakes can detach their tail if it is grasped, although this is not as widespread, nor as well developed, as it is in some lizards, whose caudal vertebrae have fracture planes that part easily. Instead, snakes may spin rapidly, thereby causing their tails to break off. Species known to autotomize their tails in this way include the Colombian Long-tailed Snake, *Enuliophis sclateri*, and the African marsh snakes, genus *Natriciteres*.

↙ Many snakes, like this Santa Cruz Garter Snake, *Thamnophis atratus*, have bold longitudinal stripes that make them difficult to keep track of when they are moving quickly, especially through vegetation.

→ A Rubber Boa, *Charina bottae*, in defensive posture, with its tail raised to imitate a false head and so deflect attack away from the real head.

↓ This Colombian Long-tailed Snake, *Enuliophis sclateri*, has lost the tip of its tail, probably due to a predator attack. This species, and a few others, have a tendency to autotomize the tail if it is grasped.

CRYPSIS

Most snakes are colored to match their background, allowing them to go unnoticed. Cryptic coloration is even more effective if a pattern or markings breaks up the snake's outline or shape, which is why most snakes have markings of blotches, spots, or bars on their dorsal surface.

Terrestrial snakes occur in a variety of colors, depending on the substrate on which they live. Forest species are usually some shade of brown to match the dead leaves, whereas desert species may be brown but may also be yellow, gray, or even pink, depending on the type of rock or sand on which they live.

← A Horned Adder, *Bitis caudalis*, camouflaged on the red sands of Namaqualand.

↙ Southwestern Speckled Rattlesnake, *Crotalus mitchellii pyrrhus*, well camouflaged when resting on the crumbled granite substrate of Joshua Tree National Park, California.

→ A gopher snake, *Pituophis* species, resting on leaf-litter, where its outline is well disguised.

Where a species ranges widely, localized color forms often crop up, corresponding with differing substrates.

Arboreal species, or snakes that live among grasses and reeds, are usually green but can also be brown or gray. Longitudinal darker stripes along the flanks are common in snakes that rely on this type of camouflage. The Central American Eyelash Pit Viper, *Bothriechis nigroadspersus*, occurs in a variety of colors and patterns, one of which is a good match for the lichens and mosses that grow on the trees and bushes where it lives, while another form is bright yellow, thought to imitate flowers and fruits of forest plants.

Disruptive coloration is especially common among vipers. This may take the form of dorsal zigzags or geometrical patterns consisting of blocks of colors, which do not necessarily match the substrate on which they live, but make the outline of the snake hard to detect. These snakes usually freeze when they detect danger, relying heavily on their camouflage for protection. Many of the well-camouflaged arboreal vine snakes do the same thing.

Fighting back

In contrast to snakes that prefer to conceal themselves from potential enemies, some species take the opposite approach. They make themselves conspicuous, and in some cases even turn the tables on an attacker.

WARNING COLORATION AND MIMICRY

Warning coloration is universal across the animal kingdom, and a wide range of species—including wasps, spiders, and amphibians—use bright colors to warn of toxins contained in their bodies and so to repel predators.

Among snakes, warning coloration is best known in the elapids, and the brightly colored species, of which there are many, are often known as "coral snakes." Coral snakes of various kinds are found in the Americas (*Micrurus* species and *Micruroides euryxanthus*), Africa (*Aspidelaps lubricus*), Asia (*Calliophis* species, *Sinomicrurus*

species, and others), and Australia (*Brachyurophis australis*). These snakes all have some areas of red, combined strikingly with black, white, yellow, or blue bands, stripes, or saddles.

There are several theories regarding the purpose of these patterns. Perhaps the most convincing is innate aversion, in which predators are "hard-wired" to avoid brightly colored animals and objects. Furthermore, the bright colors give potential predators advance warning of something distasteful or harmful,

← The Variable Coral Snake, *Micrurus diastema*, occurs in Central America, where several non-venomous snakes, such as Tschudi's False Coral Snake (above right) imitate its coloration.

↑ Bright warning coloration, and the aggressive rearing of the front half of its body, advertise the fact that this African Coral Snake, *Aspidelaps lubricus*, is venomous.

↗ Not venomous! The South American Tschudi's False Coral Snake, *Oxyrhopus melanogenys*, is a convincing mimic of the many venomous coral snakes in South America.

thereby allowing plenty of time for them to mentally process and reject it. Humans use these same bright markings on trucks, railway engines, and elsewhere for exactly the same reason.

As well as venomous coral snakes there are also "false" coral snakes: species that imitate the venomous ones despite being harmless themselves. This may be an example of Batesian mimicry, harmless species imitating harmful ones for protection. The harmless mimics, however, often live in areas where the venomous models do not occur, throwing some doubt on that particular theory, at least for some of the examples. Possibly, innate aversion and Batesian mimicry are at play in different scenarios.

Brightly colored false coral snakes are widespread. The best-known examples are the kingsnakes and milksnakes, *Lampropeltis* species, several small ground snakes, *Sonora* and *Chionactis*, from North and Central America; *Erythrolamprus mimus* and *Anilius scytale* in South America; and some pipe snakes, *Cylindrophis*, in Asia.

THREATS AND VISUAL INTIMIDATION

Snakes that feel threatened often try to avoid physical confrontation by making themselves look intimidating. This can take the form of puffing themselves up, spreading a hood and raising up the head and front part of the body, all intended to make themselves look bigger and fiercer. Striking is often a last resort, even in venomous species. Venom is a valuable resource, best saved for hunting.

Many snakes inflate their bodies, by spreading their ribs and drawing air into their elongated lungs, to make themselves look bigger than they are, including the well-named Puff Adder, *Bitis arietans*, as well as other *Bitis* species. The American hognose snakes, *Heterodon*, are known as "puffing adders" in places where they occur.

↖ An African Twig Snake, *Thelotornis capensis*, puffs up its throat when confronted to display its bold markings and make itself look larger. This back-fanged species has been responsible for human deaths.

↑ The "spectacle" marking on the hood of this Common or Spectacled Cobra, *Naja naja*, is intended to intimidate.

↗ The False Habu, *Pseudagkistrodon rudis*, from Taiwan and China, is thought to mimic the venomous Habu, *Protobothrops flavoviridis*, by spreading its jaws to emphasize its broad head and display the orange markings.

Cobras and related species are famous for raising their head and spreading a hood, which sometimes has bold patterns, such as a white circle or "spectacle" marking. Asian members of the genus *Naja*, famously used by snake charmers, are best known for hood-spreading, but similar behavior is also found in the African *Naja* species, in the two *Aspidelaps* species, and in the Rinkhals, *Hemachatus haemachatus*. The hood is spread by erecting specialized elongated ribs in the neck region. Significantly, none of these snakes rear up when hunting; hood spreading is purely a defensive tactic designed to intimidate.

Other snakes that have similar strategies, such as the Montpellier Snake, *Malpolon monspessulanus* (found in Europe, North Africa, and parts of the Middle East), do not rear up but lift their head slightly while spreading a narrow hood. The African twig snakes, *Thelotornis*, inflate their throats, also exposing bold markings. Quite a number of unrelated species, such as the Mangrove Snake, *Boiga dendrophila*, flatten their heads, sometimes exposing brightly colored labial (lip) scales. The harmless False Habu, *Pseudagkistrodon rudis*, is thought to make its head broader in this way to imitate a venomous pit viper, of which there are several from the same region.

Another strategy is "gaping." By opening its mouth widely the snake exposes the interior of its mouth, which may contrast in color with the rest of the snake. The Cottonmouth, *Agkistrodon piscivorous*, gets its name from this tendency, which it demonstrates at every opportunity.

AN AUDIBLE WARNING

Puffing up the body to make it look bigger is often a prelude to hissing. The Puff Adder, mentioned above, produces a very loud and sustained hissing sound as it deflates its puffed-up body.

Many snakes hiss in defense. The gopher and pine snakes, *Pituophis* species, have unique structures in front of their glottis that produce an impressively loud and intimidating "throaty" hiss. Other snakes hiss by exhaling or inhaling air through their nostrils, so that they can hiss without opening their mouth, and a few, notably the King Cobra, *Ophiophagus hannah*, can growl by passing air over a cavity in their throat that acts as a resonating chamber.

The rattlesnakes produce their rattling, or buzzing, sound by vibrating their tail, causing the segments of the rattle to clatter together rapidly. As the rattle

"winds down," the sound changes to a slow ticking. Some small species make a sound that is similar to that of stridulating insects. Many other snakes vibrate their tails when they are agitated, and if they are resting among dead leaves, for instance, this will produce a sound not unlike a rattlesnake.

↑   The Bull Snake, *Pituophis catenifer sayi*, rears up and produces a loud and intimidating hiss when approached. It may be mimicking venomous rattlesnakes from the same part of North America.

→   A Mojave Rattlesnake, *Crotalus scutulatus*, in full intimidation mode, with its head high off the ground ready to strike, and the rattle raised and vibrating.

Certain desert vipers have modified scales on their flanks, in which the keels are aligned obliquely and have serrated edges. When one set of scales is rubbed against another they produce a loud rasping sound (like rubbing two pieces of sandpaper together). To achieve this the snake forms a characteristic horseshoe-shaped coil and moves the opposed sections of its body against each other rapidly. A number of species use this method of intimidation, mostly venomous desert vipers, but also egg-eating snakes, *Dasypeltis*, which are in fact harmless mimics.

ODOR PRODUCTION

As anyone who has picked up a wild American ratsnake or milksnake, or a European grass snake, will tell you, some snakes are capable of producing copious amounts of foul-smelling fluid from their anal glands. This is often enough to persuade a human to release the snake forthwith (as I can personally attest).

The American fox snakes, *Pantherophis ramspotti* and *P. vulpinus*, are named for their particular brand

of repellent, and these species, as well as many others, are particularly adept when it comes to wrapping themselves around a human hand or arm while smearing as much of the musk on their captor as possible. Presumably, predatory birds and mammals are treated in a similar fashion, causing them to reconsider their next meal.

↑   The Desert Horned Viper, *Cerastes cerastes*, is a common and widespread North African species that uses its specialized scales to produce a warning sound. (Inset) The obliquely aligned keels on the scales, which the snake rubs together to produce a rasping sound, can clearly be seen in this close-up.

↗→   Southern Hognose Snakes, *Heterodon simus,* (top) are adept at pretending to be dead. Presumably, some of their predators are reluctant to eat carrion. A Barred Grass Snake, *Natrix helvetica*, (bottom) also gives the impression that it is dead by turning onto its back and allowing its tongue to hang out of its mouth. It will maintain this position for several minutes.

THANATOSIS

Related to odor production is playing dead, or thanatosis. This strategy is widespread throughout the animal kingdom so, although it may seem counterintuitive, we have to assume it is effective, otherwise it would not have evolved. Snakes that play dead include the Barred Grass Snake, *Natrix helvetica* (see page 232), which first tries to avoid capture by the use of its musk glands but then, if this fails, by flipping over onto its back and allowing its mouth to gape open and its tongue to loll out. Coupled to the distinctive and appalling smell it produces, this may persuade a predator that it has been dead for some time and is therefore inedible. For reasons that are not clear, not all Grass Snakes play dead; snakes in some populations are more inclined to do so than others.

Thanatosis as a means of defense is also used by the Rinkhals or Ring-necked Spitting Cobra, *Hemachatus haemachatus*, in South Africa, and by American hognose snakes, *Heterodon*, especially the Eastern species, *H. platirhinos*.

Venom as defense

Venom is modified saliva, and undoubtedly evolved as a means of subduing prey and beginning the process of digestion. Its use in defense is secondary, but very effective.

Members of several families, but most notably the Elapidae, Viperidae, and a small number of colubrids, produce powerful venoms that may be used in defense. These species typically warn of their ability to inflict severe pain, or death, by means of bright coloration, hissing, puffing, and so on; using up valuable venom to repel enemies is a last resort. Having said that, certain species are known for being irascible and more inclined to bite than others. Indeed, some snakes are so belligerent that they actually "jump" toward their aggressor as the momentum of their strike propels them forward.

The members of a group of cobras consisting of the Rinkhals, *Hemachatus haemachatus*, seven species of spitting cobras belonging to the genus *Naja* from Africa, and another six *Naja* species in Asia, can force venom through an aperture in the front of the fangs so that a spray is directed forward, aimed at the antagonist's eyes, where it causes immediate and intense pain. If not treated, this can cause temporary or even permanent blindness in humans.

Spitting, or spraying, is only used in defense. When they are hunting, these species strike and bite in the usual manner. Spitting may have evolved in response to the threat of humans, and is thought to have arisen more than once in different branches of the cobra family.

→ A spitting cobra rears up, opens its mouth, and sprays copious amounts of venom at an attacker, aiming at the eyes, where it causes immediate and intense pain.

BITIS GABONICA

BITIS GABONICA

Gaboon Viper

Hidden in plain sight

SCIENTIFIC NAME	*Bitis gabonica* (Duméril, Bibron & Duméril, 1854)
FAMILY	Viperidae
SUBFAMILY	Viperinae
SIZE	Up to 4 ft (1.2 m), occasionally longer, and very bulky
REPRODUCTION	Oviparous
HABITAT	Forest edges and clearings

When seen in isolation, the markings on this snake are almost unbelievably colorful, but when seen against a natural background of dead leaves its outline seems to disappear. This is a classic example of disruptive coloration, whereby a seemingly random arrangement of shapes, sizes, and colors breaks up the animal's outline.

This wonderfully patterned snake displays a geometric arrangement of rectangles and triangles colored in pastel hues of purple, pale brown, and dull pink. A row of buff markings runs along the center of its back, and the top of its head is also buff. Two bold, dark brown triangular marks radiate from the eye to the lower jaw, disguising the outline of its head and position of the eyes.

Gaboon Vipers, also known as Gaboon Adders, are one of the two largest Old World vipers, marginally smaller than the very similar West African Gaboon Viper, *Bitis rhinoceros*. Their fangs are longer than those of any other venomous snake, at a maximum of 1½ in (4 cm), and their venom is delivered in large amounts, deep into the tissue.

They are ambush hunters, waiting patiently for something edible to pass by. Their strike, and the immediate return to their resting position, is so fast that they sometimes appear not have struck at all, and the first sign that a successful strike has occurred is when the prey begins to stagger around as the venom takes effect. The snake uses its tongue to follow the dying prey's trail before swallowing it. Because of its huge head and wide jaws it is capable of engulfing animals up to the size of small monkeys and mongooses, although rodents are the most common prey.

Bites on humans are extremely rare; it is a docile and lethargic species, very slow to anger and giving ample warning in the form of a loud hiss if it feels threatened.

→ The outline of a Gaboon Viper, seen against a background of leaf-litter, is very hard to make out, helping to prevent its detection by both predators and prey.

CALLIOPHIS BIVIRGATUS

Blue Malaysian Coral Snake

Colorful rainforest dweller

SCIENTIFIC NAME	*Calliophis bivirgatus* (Boie, 1827)
FAMILY	Elapidae
SUBFAMILY	Elapinae
SIZE	5 ft (1.5 m), occasionally longer
REPRODUCTION	Oviparous
HABITAT	Montane rainforest, 1,600–3,000 ft (500–900 m) above sea level

A spectacularly colored snake with a blue body and red head and tail. The underside is also red. It lives in dense forest at higher elevations, and has a semi-fossorial lifestyle. Its bright coloration, food, and lifestyle bear some resemblance to the better-known coral snakes from South and Central America. When disturbed, it sometimes raises and coils its tail in the same way as the American Ringneck Snake, *Diadophis punctatus* (page 220).

The *Calliophis* species are thought to be the earliest evolved members of the Elapidae (cobra family). *C. bivirgatus* and six close relatives are unique among elapids in having hugely elongated venom glands, occupying one-third of their total body length. These are thought to have evolved so that they could fit inside the snake's slender body without the need for a broad head, as in most other venomous snakes, while still holding a large amount of venom. This fits in well with their semi-burrowing habits, as it enables them to move easily through leaf-litter and soil in search of the slender snakes on which they feed, with a minimum of disturbance.

Calliophis bivirgatus produces a complex venom consisting of many proteins and enzymes, different from those of other elapids. These include a unique protein known as calliotoxin, and another found in all members of the genus, maticotoxin (*Calliophis* were originally named *Maticora*). Despite the potency of its venom it is docile and rarely bites humans; only a single fatality, in the 1950s, is known.

→ A member of the cobra family, the Blue Malaysian Coral Snake is common but secretive. Its reluctance to bite makes it less dangerous than some of its close relatives.

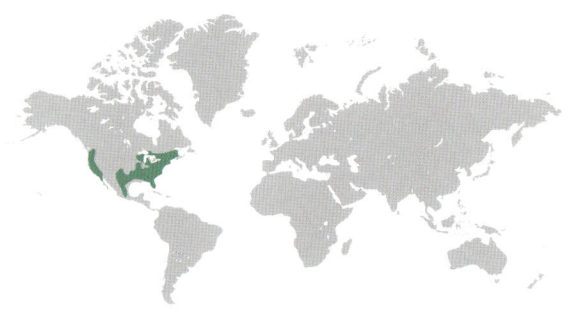

DIADOPHIS PUNCTATUS

Ringneck Snake

A secretive snake with a surprise

SCIENTIFIC NAME	*Diadophis punctatus* (Linnaeus, 1766)
FAMILY	Colubridae
SUBFAMILY	Dipsadinae
SIZE	16–20 in (40–50 cm), occasionally longer
REPRODUCTION	Oviparous
HABITAT	Woodlands, prairies, and rocky canyons

This small snake is very secretive, living beneath rocks and logs, and its distribution is dictated by the availability of these items. It also requires some moisture in the substrate, at least for part of the year, excluding it from the more arid desert areas.

Apart from these restrictions, the Ringneck Snake can be found almost anywhere within its large range, including parks, gardens, and other human-altered places, where it can be very common. Finding one of these pretty snakes while turning pieces of bark or logs is a joy.

Its diet varies from one population to another; some eat almost entirely amphibians and small reptiles, including other snakes, while others eat mainly invertebrates. It is mildly venomous and often holds prey in its mouth while chewing on it to introduce the venom; swallowing begins once the prey has stopped struggling. Due to its small size it is of no medical importance to humans.

Females produce eggs that contain well-developed embryos, which result in shorter incubation times than most other snakes; occasionally the eggs hatch almost as soon as they have been laid. Nest sites include rotting logs and piles of natural debris. Several females may use the same site, and aggregations of nearly 50 eggs have been found. A typical clutch size is 1–10.

Its defensive behavior consists of producing a foul-smelling musk and squirming, culminating in coiling the tail to display the brightly colored underside. This may be intended to startle the predator, or perhaps to divert the attack away from its head.

Twelve subspecies are recognized, varying in size and coloration, especially on the ventral surface, which may be red or yellow, plain or spotted. It seems likely that some of these will be elevated to full species at some point in the future.

→ The brightly colored underside of the tail of a Ringneck Snake is normally hidden except when it feels threatened, when it lifts its tail and forms it into a tight spiral.

Allen's Coral Snake

Multi-colored predator

SCIENTIFIC NAME	*Micrurus alleni* Schmidt, 1936
FAMILY	Elapidae
SUBFAMILY	Elapinae
SIZE	2–2½ ft (60–80 cm), occasionally to 3¼ ft (1 m)
REPRODUCTION	Oviparous
HABITAT	Lowland rainforest

There are currently 83 species in the genus *Micrurus*. Together with a single species in the genus *Micruroides*, they are collectively known as coral snakes. Most of the species in these genera are vividly marked with rings of black, white or yellow, and red, although some species only possess two of these colours. They represent the family Elapidae in the New World.

Allen's Coral Snake is a typical example, but more boldly marked than some. The markings are clearly designed to warn, and in this case they are "honest" warnings: coral snakes, especially larger ones such as this species, produce an extremely toxic venom, and bites, though rare, often result in death in humans if not treated. When handled they often thrash about, and they may bite if their mouth comes into contact with flesh. The teeth are small and the bite may not be immediately obvious; people have been bitten without realizing it until later, when the venom begins to take effect.

Allen's Coral Snake feeds mostly on a freshwater eel, *Synbranchus marmoratus*, itself an interesting fish because, although it can breathe through gills like other fish, it will crawl out onto damp ground in search of prey and breathe through the lining of its mouth and throat. When on land it is vulnerable to predation by the coral snake. Other prey animals include caecilians (legless burrowing amphibians restricted to the tropics), smaller snakes, and burrowing lizards, all of which it pursues along their burrows.

The populations on the Atlantic side of Central America have white rings, whereas those on the Pacific side have yellow rings. In this population, which has no contact with the Atlantic population, the red areas darken with age until the adult snakes are black and yellow. A number of harmless colubrid snakes have similar markings, thought to be examples of Batesian mimicry (see page 207).

→ One of the more dangerous South American elapids, Allen's Coral Snake has an almost cylindrical body and narrow head, allowing it to travel along burrows in search of prey.

Atlantic Central American Milksnake

Multi-colored coral snake look-alike

SCIENTIFIC NAME	*Lampropeltis polyzona* Cope, 1860
FAMILY	Colubridae
SUBFAMILY	Colubrinae
SIZE	2–3¼ ft (60–100 cm), occasionally longer
REPRODUCTION	Oviparous
HABITAT	Tropical and subtropical deciduous forests, dry hillsides and farmland

This is one of several coral snake mimics, species that are thought to gain an advantage by looking similar to one of the highly venomous coral snakes, *Micrurus* species (see pages 207 and 222). They are known collectively as "false coral snakes."

The Atlantic Central American Milksnake and related species rely on their brightly contrasting pattern to confuse possible predators, gaining them precious time to make an escape. Mostly nocturnal, these snakes hide by day under rocks or logs or in burrows. If they are unearthed by a predator, they will often move quickly and erratically, and may make false strikes, to further mimic the venomous species. This is often proposed as an example of Batesian mimicry, a widely accepted theory that explains why some harmless species look similar to some dangerous ones (see page 207).

Other experts, however, have pointed out that venomous coral snakes are so lethal that predators attacking them would probably be killed and would therefore be unable to learn to avoid them. They propose an alternative theory, known as "innate aversion," which relies on the fact that animals are instinctively reluctant to attack brightly colored snakes (and insects such as wasps and bees), and this explains the bright colors of both lethal and harmless species.

Apart from several *Lampropeltis* species, false coral snakes are also represented in several other genera, including the Coral Cylinder Snake, *Anilius scytale* (which is the only member of its family, Aniliidae), the Centipede Snake, *Scolecophis atrocinctus*, and several members of the genus *Oxyrhopus*.

There is an often-quoted rhyme which goes "Black on yellow, kill a fellow; red on black, venom lack." Unfortunately, this rule only applies to snakes in the United States. Further south there are so many variations in the color patterns of coral snakes that it is not possible to generalize in this way.

→ The Atlantic Central American Milksnake, *Lampropeltis polyzona*, has been the subject of many name changes over recent years. It is now divided into several subspecies, all of which are thought to be mimics of the similar, but venomous, coral snakes.

SCIENTIFIC NAME	*Naja mossambica* Peters, 1854
FAMILY	Elapidae
SUBFAMILY	Elapinae
SIZE	2½–4¼ ft (80–130 cm)
REPRODUCTION	Oviparous
HABITAT	Grassland and woodland, often near water but also in drier places

NAJA MOSSAMBICA

Mozambique Spitting Cobra

Defense at a distance

One of the smallest African cobras and often overlooked, although it may be the most common snake in some areas, even entering houses in search of prey. Despite its small size this species can be dangerous to humans.

All spitting cobras, of which there are eight species in Africa and two in Asia, have modified fangs, with the opening positioned in the front of the fang rather than the tip. When venom is forced out under pressure it forms a fine spray that may reach more than 6 ft (2 m). Spitting cobras aim at their antagonist's eyes; the effects can be instantly painful and may lead to permanent blindness. First aid consists of flushing the eye out with large quantities of water or milk to dilute the venom.

The Mozambique Spitting Cobra is very quick to spit, and this, coupled with its inclination to enter villages and houses, results in many incidents. It also bites without spitting, and is thought to cause the greatest number of venomous snake bites in the parts of southern Africa where it is found. Bites are rarely fatal but may cause extensive tissue damage.

Its defensive behavior consists of raising the head and neck off the ground and spreading a hood by moving a number of elongated ribs in its neck outward, causing loose skin to stretch between them. The markings on its neck are often black and white bars or blotches, so that it is more conspicuous, while the hood makes the snake look bigger than it is. Spitting cobras do not spread their hoods, or spit, when they are hunting.

→ The defensive action of a Mozambique Spitting Cobra involves raising its head, spreading a narrow hood, and squirting venom for distances of 7 ft (2 m) or more.

LACHESIS STENOPHRYS

Central American Bushmaster

The largest pit viper

SCIENTIFIC NAME	*Lachesis stenophrys* Cope, 1875
FAMILY	Viperidae
SUBFAMILY	Crotalinae
SIZE	6½ ft (2 m), exceptionally longer
REPRODUCTION	Oviparous
HABITAT	Lowland rainforest

The largest viper in the Americas and the longest in the world, the Bushmaster reportedly grows to nearly 13 ft (4 m), possibly more. A large Bushmaster in the wild is an awe-inspiring sight, especially when encountered at night.

Bushmasters are covered in small, rough scales that have raised, knobbly keels, resembling the overall texture and color of a pineapple, a fact that is referred to in several local names. Its tail ends in a long spine-like scale, which the snake vibrates if disturbed; another local name for the species is *Cascabela Muda*, or "Mute Rattlesnake." This huge viper is restricted to primary forests and is rarely found where the habitat has been altered for agriculture, although it may linger on in freshly cleared areas. This brings it into contact with humans, and it is invariably killed when this happens.

It is not an aggressive snake but very alert, and bites are not unknown. They are usually fatal if not treated, or even when they are treated; records from Costa Rica showed an 80 percent death rate.

They are ambush predators, and their main prey is rodents. They will station themselves next to a rodent trail, often beneath a fruiting tree, and return to the same place every night until they are successful. After feeding, they move to a hollow log or similar retreat and remain there until they have digested their meal.

Bushmasters are the only oviparous New World pit viper. A clutch of 5–19 eggs is laid in a disused mammal burrow or other underground chamber, and the female remains with them during the incubation period, which lasts 9–11 weeks. The young are 12–20 in (30–50 cm) long when they hatch.

→ The Central American Bushmaster is a very large, impressive pit viper. It is greatly feared, but rarely implicated in snakebites, which are normally the result of confrontations with smaller, more common, species.

West African Saw-scaled Viper

A viper with rasping scales

SCIENTIFIC NAME	*Echis ocellatus* Stemmler, 1970
FAMILY	Viperidae
SUBFAMILY	Viperinae
SIZE	12–20 in (30–50 cm), occasionally longer
REPRODUCTION	Oviparous
HABITAT	Savanna and sparse woodland

The intricate pattern on the back of this snake, and the other 11 *Echis* species, is the source of their alternative name of carpet vipers. The West African species, described only in 1970, is the main cause of venomous snake-bite deaths in Senegal and a major cause of bites in neighboring countries. Rural workers using short-handled tools and working barefoot is a significant factor.

The West African Saw-scaled Viper is almost completely terrestrial, only occasionally climbing into bushes, and is nocturnal. During the day it shelters in holes, under rocks, or in piles of brush. Its camouflage is most effective at dusk when it ventures out to hunt for food, and this is the time when it is most likely to come into contact with workers returning from their fields. Bites cause local swelling, and copious bleeding due to its anticoagulant properties. The venom is relatively slow-acting in humans, and if medical facilities are available there is usually time for treatment to prevent a fatality.

Its defensive behavior consists of forming a number of C-shaped coils which are moved against each other to produce a surprisingly loud rasping or hissing sound. At the same time, the snake strikes aggressively, sometimes moving forward with the momentum of the strike. The harmless egg-eating snakes, *Dasypeltis* species, mimic the threat display of saw-scaled vipers and have the same arrangement of sound-producing oblique, serrated keels on the scales on their flanks.

→ An irascible hazard for rural workers in parts of West Africa, the West African Saw-scaled Viper is one of several desert snakes that produce a warning sound from their modified scales.

Barred Grass Snake

The swimmer

SCIENTIFIC NAME	*Natrix helvetica* (Lacépède, 1789)
FAMILY	Colubridae
SUBFAMILY	Natricinae
SIZE	4 ft (1.2 m), but occasionally to 6½ ft (2 m); females much larger than males
REPRODUCTION	Oviparous
HABITAT	Damp places, including marshes, ponds, and lakes; occasionally woods

The Barred Grass Snake is one of only three British snakes, and the only egg-laying species. It occurs as far north as the Scottish border and there are single records for Ross-shire and Inverness-shire. It also occurs in parts of western Europe. Until recently it was lumped together with two other "grass snakes," *N. astretophora* from the Iberian peninsula, and *N. natrix* from eastern Europe.

The Barred Grass Snake is a fast-moving, semi-aquatic snake usually found close to water but which may also stray into dry, grassy areas. It eats mainly amphibians and fish and is one of the few species that will eat toads, despite the poison they secrete from glands in their skin; in drier places toads are probably their main food.

Their most important adaptation to cold is their choice of where they lay their eggs. Under normal circumstances, the eggs would fail to hatch were it not for the mother's ability to seek out warm places to lay them. These include piles of rotting vegetation such as leaves, compost and manure heaps, piles of rotting seaweed, and, in the past, even holes in the brick walls of rural baking ovens, when such things existed. Several females will often use the same site, resulting in 100 or more eggs incubating at the same site.

The Barred Grass Snake is well known for its defensive strategy of feigning death, by releasing a foul-smelling fluid from its cloaca, turning onto its back and allowing its tongue to hang out of its mouth. This apparently deters predators, who may be reluctant to eat a dead, apparently decomposing, carcass. Despite this well-recorded behavior, it is not that common; of many dozens of Barred Grass Snakes I have found in England and Wales, none has ever performed for me.

→ The most common snake over much of its range, the harmless Barred Grass Snake can be found in urban areas, visiting garden ponds and laying its eggs in compost or dung heaps.

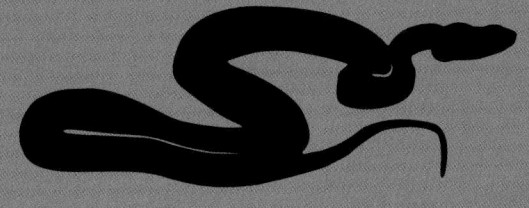

SNAKES AND HUMANS

An uneasy relationship

In the West, snakes and humans have had an uneasy relationship ever since Eve was tempted by the serpent in the Garden of Eden. Other cultures have less entrenched attitudes, but there is still a feeling in many places that "the only good snake is a dead snake."

Much of this attitude arises through ignorance. Studies have shown that we empathize most with animals that look like us. We prefer animals with four limbs, an upright stance, and eyes directed forward, such as monkeys, Giant Pandas, and bears. Snakes don't come out of this comparison well.

Snakes can appear and disappear quickly and mysteriously, they crawl on their stomachs, and they cover the ground with no visible means of propulsion. These characteristics generate fear as well as fascination. On top of this, some of them are able to deliver pain or even death in the blink of an eye.

The truth of the matter is that, although snakes cause fatalities in parts of the world where they come into close contact with humans, the chances of being killed by a snake are vanishingly small in most places owing to their timid natures and relative scarcity.

In recent years a more enlightened attitude seems to have taken root, at least among the more rational sections of the population. Eco-tourism has opened up the world to interested travelers, and television programs have heightened the public's awareness of nature as a whole, emphasizing the importance of preserving entire plant and animal communities. Snakes are not a thing apart but an important component in a complex web of life. As the environmentalist Aldo Leopold famously wrote, "The first rule of intelligent tinkering is to keep all the parts."

↙   A belt buckle in the shape of a bronze snake. These are regularly found by metal detectorists although dating them can be problematical.

→↘↓   Works of art from various cultures sometimes depict snakes in some form or another, evidence of the importance that has been attached to them over the ages.

Snakes in culture and mythology

Snakes have a long association with humans going back to prehistorical times and continuing right up to the present day. Human–snake associations, encompassing religion, medicine, and fertility, exist in many parts of the world and among many cultures. There is space to mention only a few of them here.

Snakes are mysterious, which is always a good reason to worship or fear something. They can appear unexpectedly, having spent a period of time in underground chambers or crevices to avoid heat, drought, or freezing conditions. This often coincides with springtime, or a rainy season, both of which are intimately connected to the rhythm of human activities, especially agriculture and hunting.

→ Petroglyphs from Utah show shapes that look like snakes, alongside other animals that were obviously important to the local inhabitants.

← To members of the Hopi tribe in North America, snakes were messengers from the underworld and figured in many of their dances and festivals.

→ Stylized snakes can be found nowadays in many of the emblems associated with medical and pharmaceutical organizations.

↘ The "Asp" used by Cleopatra to kill herself was almost certainly a cobra, if indeed she did die of snakebite, because the true Asp Viper, *Vipera aspis*, does not occur in Egypt.

A SYMBOLIC SIGNIFICANCE

The earliest indications that humans interacted with snakes come from cave paintings in southwestern Europe, southern Africa, and Australia. They depict snakes or snake-like creatures. Snakes may have been important because they were a source of food, but, equally likely, they may have had a more symbolic significance.

Australian aboriginal mythology, for instance, includes the story of rainbow serpents, which lived hidden in deep waterholes during the dry season but manifested themselves as rainbows during the rainy season; they were the custodians of water supplies, vitally important to a society that lived in an arid environment.

Similarly, the Hopi people of North America have, for thousands of years, used snakes in their ceremonies, believing that they have a connection to the underworld, linking it to the world above and bringing fertility to their fields through rain and storms.

Venomous and non-venomous species figure in their annual snake dance, and act as messengers to the gods when they are released into cracks and crevices in the ground.

SACRED SNAKES

The ability of snakes to shed their skin, ridding themselves of an old, dull, and worn integument and replacing it with a fresh one, can be likened to a rebirth and the cycle of life.

Snakes were sacred to the ancient Greeks, and were associated with the god Asklepios, who was later incorporated into the Roman mythological world as Aesculapius. This is the serpent we see today featured in the emblem of the medical profession, coiled around a staff. It is probably the species *Zamenis longissimus*, known as the Aesculapian Snake. It is possible that snakes of this species were introduced to temples and other sacred places throughout Europe, which would account for its disjunct distribution (see page 274).

In Hinduism and Buddhism, cobras are worshiped in the form of the Nagas (hence the scientific name of the genus, *Naja*) for their association with rebirth. In China the dragon serves the same purpose, and is also believed to control rainfall and other phenomena. Similarly, the Egyptians believed that snakes were associated with rain, the River Nile, and therefore fertility. A rearing cobra known as the Ouraeus was incorporated into the headdress of the Pharaoh and also used as a hieroglyph depicting a goddess.

The Christian religion has not followed the trend of venerating snakes, perhaps for the very reason that the snake was associated with perceived heathen cultures. Indeed, it epitomizes evil, responsible for Man's temptation in the Garden of Eden.

Problem snakes

A mention of "problem snakes" may first prompt thoughts of human snakebite fatalities, but snakes can also cause problems for other wildlife. In a few cases, invasive introduced species have given rise to serious conservation concern.

BROWN TREE SNAKES

In the 1940s a population of Brown Tree Snakes, *Boiga irregularis*, was established on the Pacific island of Guam, having arrived via cargo delivered to the United States military base. It had no natural predators on the island and spread rapidly. At present there are reckoned to be two million of them on Guam.

The Brown Tree Snake is a generalist, eating whatever it can catch (see page 268). The native birds on Guam, never having encountered snakes before, had little defense against it and they were decimated within 20 years. Several species were completely exterminated, including three endemics, the Guam Flycatcher, *Myiagra freycineti*, the Guam Rail, *Gallirallus owstoni*, and the Guam Kingfisher, *Todiramphus cinnamominus*. The latter two species have been established in captivity

in an attempt to save them from extinction, but the Guam Flycatcher appears to have gone for good. Attempts to eliminate the snake have not proved successful, although over $4m has been set aside recently to devise a program to control them.

BURMESE PYTHONS

Meanwhile, in Florida, Burmese Pythons, *Python bivittatus*, have established themselves in the area around the Everglades. They are assumed to have originated from pet snakes that escaped or were deliberately released. However they got there, the situation is now completely out of hand, with a population currently estimated at 150,000 and growing. One of the larger individuals caught so far measured 17 ft (5.2 m). More worryingly, it contained 122 eggs. A clutch found in 2023 contained 111 eggs (see also page 272).

These large snakes have equally large appetites, and so far 24 species of mammals, 47 species of birds, and two species of reptiles have been found inside their stomachs. There are campaigns to eliminate them or reduce their numbers, but it may already be too late.

← The Brown Tree Snake, *Boiga irregularis*, is the cause of much damage.

↑ Burmese Pythons, *Python bivittatus*, are now common in the Florida Everglades, an area that clearly suits them too well. Attempts to eliminate them are almost certainly doomed to failure.

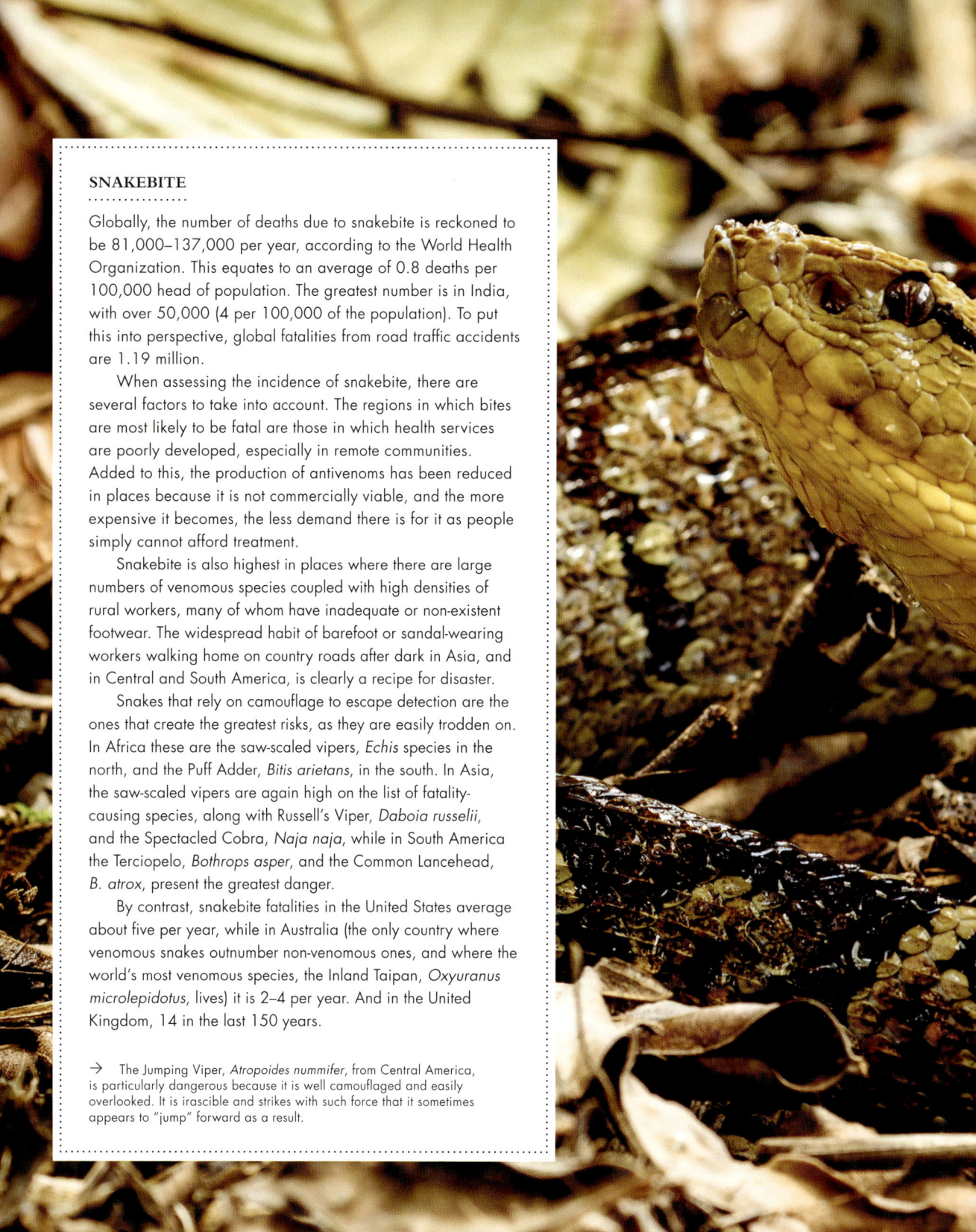

SNAKEBITE

Globally, the number of deaths due to snakebite is reckoned to be 81,000–137,000 per year, according to the World Health Organization. This equates to an average of 0.8 deaths per 100,000 head of population. The greatest number is in India, with over 50,000 (4 per 100,000 of the population). To put this into perspective, global fatalities from road traffic accidents are 1.19 million.

When assessing the incidence of snakebite, there are several factors to take into account. The regions in which bites are most likely to be fatal are those in which health services are poorly developed, especially in remote communities. Added to this, the production of antivenoms has been reduced in places because it is not commercially viable, and the more expensive it becomes, the less demand there is for it as people simply cannot afford treatment.

Snakebite is also highest in places where there are large numbers of venomous species coupled with high densities of rural workers, many of whom have inadequate or non-existent footwear. The widespread habit of barefoot or sandal-wearing workers walking home on country roads after dark in Asia, and in Central and South America, is clearly a recipe for disaster.

Snakes that rely on camouflage to escape detection are the ones that create the greatest risks, as they are easily trodden on. In Africa these are the saw-scaled vipers, *Echis* species in the north, and the Puff Adder, *Bitis arietans*, in the south. In Asia, the saw-scaled vipers are again high on the list of fatality-causing species, along with Russell's Viper, *Daboia russelii*, and the Spectacled Cobra, *Naja naja*, while in South America the Terciopelo, *Bothrops asper*, and the Common Lancehead, *B. atrox*, present the greatest danger.

By contrast, snakebite fatalities in the United States average about five per year, while in Australia (the only country where venomous snakes outnumber non-venomous ones, and where the world's most venomous species, the Inland Taipan, *Oxyuranus microlepidotus*, lives) it is 2–4 per year. And in the United Kingdom, 14 in the last 150 years.

→ The Jumping Viper, *Atropoides nummifer*, from Central America, is particularly dangerous because it is well camouflaged and easily overlooked. It is irascible and strikes with such force that it sometimes appears to "jump" forward as a result.

Threats to snakes

Throughout the world, snakes face many threats from humans. Habitat destruction for agriculture and development is probably the greatest of these, and the biggest cause of species extinction, but snakes are also under attack from many other sides, including the introduction of invasive species, the road-building that accompanies development, and widespread prejudice.

↓ Rainforest in Ranomafana National Park, Madagascar, an area that has been protected since 1991 when rare lemurs were discovered here. It is equally important for many reptiles, including snakes.

→ By contrast with the photograph below, this area in central Madagascar has no protection, and generations of slash-and-burn agriculture have resulted in a severely degraded habitat.

HABITAT DESTRUCTION

Although no wildlife is immune, animals such as snakes, which have poor powers of relocation, are most seriously affected by the destruction of habitat wrought by humans.

Habitat destruction goes hand-in-hand with road-building and other forms of habitat fragmentation restricting the movement of animals from one part of their range to another. Snakes that are adaptable, especially small species that can find a niche in urban habitats, may linger for a while, but local extinction is usually inevitable. Fragmented habitats reduce genetic diversity, and inbreeding may result.

The regions with the greatest diversity of snake species are all in the tropics: Central and South America, especially the Amazon Basin; Southeast Asia; and sub-Saharan Africa. These are also the regions currently most affected by habitat destruction, mainly for agriculture and for the plundering of natural resources. Almost half the world's tropical rainforests have been destroyed already, and destruction continues apace. Many snakes living in these regions are yet to be studied, or even described, and some will go extinct before they are known to exist.

Away from the rainforests, dry tropical forests have been destroyed through "slash and burn" clearance and agriculture. In Madagascar, for instance, less than 8 percent of the original forest remains.

Other snake habitats have been affected, though to a lesser degree. Wetlands have been drained, grasslands and scrub have been put to the plow, and irrigation schemes have been built to enable deserts and semi-deserts to be converted to agriculture, much of it short-term.

HARMFUL INTRODUCTIONS

In many parts of the world where snake populations are declining, the second-biggest threat, after humans themselves, is posed by alien species. These may be introduced either intentionally or accidentally, and the effects can be devastating.

In Australia, Cane Toads, *Rhinella marina*, were introduced in 1935 in a misguided attempt by the sugar industry to control cane beetles. The toads thrived

(as did the beetles) and spread throughout Queensland and into neighboring states. They are still spreading. The toads had a dramatic and negative effect on snakes, many of which ate native frogs and had no immunity to the powerful toxins produced by Cane Toads. Studies have shown that at least 49 species of snakes have been affected.

As the toads spread across the country they will encounter new snakes, which will also be impacted. In Queensland, populations of the Northern Death Adder, *Acanthophis praelongus*, declined by 89 percent in regions where the toads had spread. But the Red-bellied Black Snake, *Pseudechis porphyriacus*, which was also badly affected, seems to have become partially immune to the toads' toxins, and has also developed an aversion to eating them.

In another misguided attempt at biological control, the Small Indian Mongoose, *Urva auropunctata*, was introduced to many islands, especially in the Caribbean region, in order to reduce rat populations. They are highly predatory on snakes and have wiped out several island populations of racers, *Alsophis* species, and

brought others to the brink of extinction. Now they are present on more than 60 islands around the world, and they are still spreading.

All these introductions pale into insignificance, though, when compared with domestic and feral cats. In Australia alone cats kill an estimated 1.6 million reptiles *per day* or 609 million per year. Although the majority of these are lizards, the effects on snake populations must be enormous (and, since many snakes eat lizards, this will have an indirect effect on them). In the United States cats are believed to kill between 228 and 871 million reptiles each year.

↖ The introduction of the Cane Toad, *Rhinella marina*, to Australia, with its massive poison-secreting glands, has wrought havoc with native predators, including several species of snakes.

↑ A Javan Mongoose, *Herpestes javanicus*, confronting an Indonesian Cobra, *Naja sputatrix*.

↗ Just one example of several tens of thousands of snakes that are killed by traffic every year.

ROADKILL

Urban and agricultural development both mean more roads, and more potential hazards for the snakes that live alongside these roads.

Numerous studies in many parts of the world have been carried out to assess the impact of road traffic on snakes and other vertebrates. Snakes are especially vulnerable compared with other groups because they are relatively slow-moving and elongated, increasing the chances of them being hit (and longer snakes are more vulnerable than shorter species). During the early part of the night, nocturnal snakes often linger on the surface of tarred roads, basking in the warmth retained there, increasing the chances that they will be run over.

Accurate figures are hard to obtain. Studies can only monitor relatively short stretches of roads for short periods of time, and the majority of casualties are taken by scavengers before they can be recorded. So the true figures are likely to be much higher than surveys would indicate. The general conclusions, however, are that roads passing through agricultural and urban areas are the most hazardous for snakes. Roads crossing nature reserves and other protected areas also account for high numbers of roadkills, presumably because snakes are more numerous in these places.

PREJUDICE

Sadly, some drivers will go out of their way to deliberately hit snakes on the road. This falls into another category of threats to snakes: prejudice through ignorance. This is only too common, but the worst examples are the American rattlesnake "roundups" in which hundreds, sometimes thousands, of rattlesnakes are captured and slaughtered for entertainment and profit. The states with the most roundups are Texas, where the Sweetwater roundup is the largest and best known, and Oklahoma.

Exploitation

There are three main ways in which snakes are exploited by people: for food, for skins, and for the pet trade. Although it is not always possible to quantify the numbers involved, because they are mostly unrecorded, the impact locally must be considerable.

Snakes are eaten in some parts of the world, notably in Southeast Asia. Large numbers of sea snakes are eaten here, along with many terrestrial species including cobras, kraits, ratsnakes, and pythons. In fact, any snake of a worthwhile size is likely to be eaten. Small numbers of snakes are also eaten in parts of rural Africa and Australia—where, again, pythons are especially favored because of their size. Rattlesnakes are eaten as a novelty food in North America, the market for them being linked to the roundups mentioned on page 249. Certain snake parts are also used in Chinese medicine.

The trade in snakeskins is concentrated on a few species, mainly the Reticulated Python, *Malayopython reticulatus*, and the Burmese Python, *Python bivittatus*.

← Snake skins are processed in large numbers, often under primitive conditions, in parts of Africa and Asia.

↑ A snake charmer in Rajasthan, India. Displays such as this are a staple ingredient on the tourist trail.

↗ A Ball or Royal Python, *Python regius*, with attractive coloration specially bred for the pet market.

Although the trade is often illegal, financial incentives are such that large numbers of snakes are killed. Annual figures for illegal trade of up to $1bn per year have been estimated in the recent past, although the official figure, for skins traded legally, is less than half this. For instance, approximately half a million skins are imported legally from Southeast Asia to Europe per year. Each skin is worth about $700, bringing the total to $350 million. Overall numbers are falling, however, as wild snakes become harder to find, although python farms have been set up in Thailand to supply the trade.

Snakes are widely kept as pets in North America and Europe. The trade in wild snakes has reduced dramatically in the last twenty years as captive breeding techniques have improved, and there is now little need to trade in wild-caught individuals, which, in any case, are often parasitized and stressed, and are therefore short-lived.

Climate change

Very little research has been done to assess the implications of climate change for snakes. Climate change will probably accelerate, and it is likely to have a bigger impact in the future than it has in the past.

Being ectotherms, snakes are very sensitive to changes in temperature patterns. Species from temperate regions may emerge earlier in the year if average temperatures rise, and may benefit from a longer active season. On the other hand, erratic warm spells during the winter can encourage snakes to emerge early and then become stranded when the weather turns cold again.

Synchronization of breeding seasons could become harder to maintain. Snakes living in cooler regions may be able to expand their ranges as the climate warms up and new areas become habitable to them, but those living in hot, arid regions may be vulnerable to overheating and drought.

Cyclones are becoming more frequent and can cause tidal surges, affecting snakes living in coastal areas and on islands, some of which will be reduced in area in the long term. Coral reefs appear to be dying due to increasing sea temperatures, and the sea snakes that rely on them for feeding will be affected, while populations of sea kraits, which require sea caves in which to lay their eggs, will suffer if rising sea levels and storm surges flood the caves with saltwater.

← Global warming and drought are precursors of habitat change, which is already impacting populations of snakes, and their effects will accelerate if they are not tackled.

↗ Forest fires that destroy large swathes of land are especially harmful to creatures such as snakes that are unable to flee. Recovery of the habitat may come too late for any that have managed to escape underground.

→ Shallow corals beginning to bleach in eastern Indonesia. Bleaching is usually caused by increasing sea temperature and involves corals' symbiotic algae being expelled. This turns the colony white.

Inland, heavy rainfall and flooding will inundate low-lying areas, affecting snake populations, especially burrowing species and snakes that have gathered underground to hibernate and which will be unable to escape to higher ground. Excessively hot weather may bring snakes, including dangerous species, into closer contact with humans as they seek to escape from the heat by sheltering in houses and outbuildings.

Extinct and endangered snakes

Many snakes are in danger of extinction, and several have already suffered this fate. Owing to their (literally) low profile, rare snakes are easily overlooked, and it is highly likely that some species have gone extinct before they could be discovered. Similarly, many more that we are unaware of are becoming scarcer.

EXTINCT SNAKES

- Hoffstetter's Worm Snake,
  Madatyphlops carieri (Mauritius)
- Round Island Burrowing Boa,
  Bolyeria multocarinata (Round Island, Mauritius)
- Barbados Ground Snake,
  Erythrolamprus perfuscus (Barbados)
- Underwood's Mussurana,
  Clelia errabunda (St. Lucia, Caribbean)

→ A preserved specimen of the Round Island Burrowing Boa, *Bolyeria multocarinata*, a species that has not been seen for many years and is presumed extinct.

The most important body monitoring and recording wildlife is the IUCN (International Union for Conservation of Nature). This is a global organization created in 1948 covering nature in over **170** countries. It assesses populations of species, grades them according to their status, and publishes its findings in the IUCN Red List.

The most recent information from the Red List includes four species of snakes that are extinct (see box above). These were all restricted to small islands. A further 85 species are classified as Critically Endangered, and 12 of these are classed as Possibly Extinct. Again, island species figure prominently in this list. Finally, **179** species are classed as Endangered, and **161** as Vulnerable. Bear in mind that these are the species we know about. There will be others that live in less-studied regions that have already gone extinct or are likely to do so in the near future.

Causes of the four known extinctions include habitat destruction due to clearance for agriculture, habitat degradation due to the introduction of goats, predation by introduced rats, cats, and mongooses, and human persecution.

Island distributions also figure prominently in those species considered to be Critically Endangered and Vulnerable. Again, introduced predators are a serious problem, and all species in these categories are affected by habitat clearance resulting from one or more of the usual reasons: agriculture, including tea, banana, and oil palm plantations, cattle grazing, timber extraction, mining, quarrying, urban expansion, and development for tourism. The future for many of these species does not look hopeful.

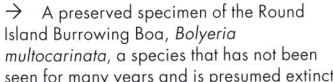

← The Milos Viper, *Macrovipera lebetina schweizeri*, occurs only on a group of small Aegean islands in Greece, where its survival is threatened by habitat change due to tourism and mining.

Threatened sea snakes: an overlooked problem

The 70 or so species of sea snakes can be divided into two groups: the live-bearing species that need never come ashore throughout their lives; and the egg-laying species (the sea kraits) that must find suitable locations along the shoreline in which to deposit their eggs. These latter species also mate and shed their skins on land, and may come ashore to drink from freshwater pools.

All marine snakes are difficult to study and poorly known compared with terrestrial species; their status is often overlooked for these reasons.

In a recent survey, two species, *Aipysurus apraefrontalis* and *A. eydouxii*, were identified as Critically Endangered, and a third, *A. fuscus*, was classified as Endangered. All three inhabit small sections of coral reefs (the Ashmore Reefs) in the Timor Sea and, although it is not clear why they are faring so badly, reef bleaching and raised sea temperatures, due to climate change, have been proposed as possible causes. Worryingly, severe coral bleaching and mass mortality are predicted for the next several years.

The sea snake *Hydrophis semperi* and the sea krait *Laticauda crockeri*, which live in large land-locked lakes in the Philippines and the Solomon Islands respectively, are considered to be Vulnerable as their habitats are becoming degraded, whilst another sea krait, *L. schistorhyncha*, from the small island of Niue in the South Pacific, which is dependent on reefs and intertidal habitats, is also Vulnerable.

A further 23 species of marine snakes are currently classified as Data Deficient, as they are poorly known, some of them from just a handful of specimens collected accidentally as fishermen's by-catch.

In summary, there are 69–71 species of sea snakes (the taxonomic status of some of them is disputed at present) and, of these, 26 species (37 percent) are a cause for concern.

← Stokes' Sea Snake, *Hydrophis stokesii*, is, at present, widespread and apparently not one of the more endangered species.

↑ Sea snakes and sea kraits lead secretive lives, unseen by the general public, but they are important elements of reef ecology. Changing climate patterns are affecting some of the reefs on which they live, resulting in the reduction in numbers of some species, a fact that has largely gone unnoticed.

↑　Yellow-lipped Sea Krait, *Laticauda colubrina*. This is the most widespread of the sea kraits.

↗　Slender-necked Sea Snake, *Hydrophis melanocephalus*, from the Indian Ocean and the South China Sea.

Conservation

Many of the threats to snakes, such as habitat destruction and climate change, are irreversible. Others, such as persecution and the introduction of alien species, can be reversed whenever there is a concerted effort to do so.

At present, conservation efforts are conducted on several fronts. Where habitat destruction is a factor, steps have been taken to protect remaining areas, involving legislation where practicable. This can go hand-in-hand with education, to help local people understand the value of the animals that live near them. Programs exist in many parts of the world, including Central and South America, Africa, and India (where the King Cobra is subject of a high-profile awareness project, for instance).

Where alien species such as goats, mongooses, rats, and cats are the main problem, eradication programs have been instigated. This has already been successful on some small islands, but eradicating alien species from mainland areas is much more difficult.

← Kamal Devkota, leader of the "Save the Lives" program in Nepal, with one of the King Cobras, *Ophiophagus hannah*, he works with.

↗ Round Island Boas, *Casarea dussumieri*, have been successfully bred at Jersey Zoo, United Kingdom, and reintroduced to the wild.

Where all else fails, or sometimes in tandem with these measures, remaining individuals are captured and brought into a captive-breeding program. The Round Island Ground Boa, *Casarea dussumieri*, for example, was the subject of a successful 30-year captive-breeding project by Jersey Zoo, with 70 captive-bred young adults eventually translocated to the island of Gunner's Quoin, which is close to their original home on Round Island, near Mauritius. Meanwhile, feral goats and rabbits were removed from Round Island to restore the original habitat. Both Round Island and Gunner's Quoin are closed reserves, with no human visitors allowed.

Similar projects are under way in the Caribbean, where mongooses are a particular additional problem, and where the Antiguan Racer, *Alsophis antiguae*, has become extinct on the main island of Antigua. Following eradication of alien species on 15 small satellite islands, the snakes have been successfully reintroduced to three of them. Unfortunately, this species has not responded well to captive-breeding attempts, a fate which is common to many snakes whose diet and environmental conditions are not easily replicated.

King Cobra

The longest venomous snake in the world

SCIENTIFIC NAME	*Ophiophagus hannah* (Cantor, 1836)
FAMILY	Elapidae
SUBFAMILY	Elapinae
SIZE	10 ft (3 m), occasionally to over 13 or even 16 ft (4–5 m)
REPRODUCTION	Oviparous
HABITAT	Primary and secondary rainforest

The King Cobra, also known as the Hamadryad, is an iconic species, an apex predator that feeds on other snakes, and the longest venomous snake in the world. It occasionally exceeds 13 ft (4 m) in length, and a large one can rear up until it is eye-to-eye with a human.

This species has complex social behavior. It is territorial, and males fight by entwining their bodies and trying to press their rival to the ground, with the dominant male earning the right to mate. Before laying, the female builds a nest out of dead leaves, using a loop of her body to gather them into a heap and then staying with the eggs until they hatch, about 70 days later. The hatchlings are brightly colored in black and white bands or chevrons and are arboreal at first. Some individuals retain the striped pattern, although it becomes less bold, while others lose it altogether to become uniform brown.

This is a species in trouble. Overall populations have fallen by at least 30 percent, even 80 percent in some parts of their large range, due to habitat destruction, persecution, and exploitation for the skin trade. It requires wet rainforests in which to live, and many such areas have been cleared for agriculture, bringing it into conflict with humans. This is not an aggressive snake, however, and bites are very few.

Moves are afoot to change the perception of these snakes in rural communities, spearheaded by local experts who relocate problem snakes that have strayed into villages, while researching their ecology and educating local people.

Its prey includes kraits, Spectacled Cobras, *Naja naja*, Russell's Vipers, *Daboia russelii*, and pit vipers—the species that are responsible for the majority of the over 50,000 snakebite deaths that occur in India annually. For this reason alone, the King Cobra is worth protecting.

→ Adult King Cobras may be plain black or brown in color, or they may retain vestiges of the banded pattern they have when they hatch.

OXYURANUS MICROLEPIDOTUS

Inland Taipan

Not so fierce after all

SCIENTIFIC NAME	*Oxyuranus microlepidotus* (McCoy, 1879)
FAMILY	Elapidae
SUBFAMILY	Hydrophiinae
SIZE	6½ ft (2 m)
REPRODUCTION	Oviparous
HABITAT	Wetlands and floodplains, associated with cracked alluvial soils

The Inland Taipan, also known as the Fierce Snake, is credited with the most potent venom of any land snake. Despite its fearsome reputation, it comes from a remote part of Australia where contact with humans is rare and bites almost unheard of.

The only recorded bites from this species all involve people who were handling captive snakes. All were treated promptly with antivenom and all eventually recovered. This is despite the fact that, theoretically at least, it can deliver enough venom to kill 100 humans and is four times as toxic as its nearest relative, the Coastal Taipan, *Oxyuranus scutellatus.*

The Inland Taipan lives in an area that is frequently flooded, leaving beds of cracked clay soil that provide shelter in the form of tunnels and underground chambers. It shares this habitat with its most important prey species, the Long-haired Rat, *Rattus villosissimus*, whose population fluctuates dramatically in response to erratic rainfall,

so the snake's fortunes are indirectly linked to climate. When its preferred prey is scarce it switches its diet to other small mammals.

It strikes rapidly and bites down repeatedly before releasing its victim. Its venom contains a cocktail of neurotoxins, including paradoxin, one of the most potent toxins known, hemotoxins, which affect the blood, and myotoxins, which affect muscle, as well as other compounds in smaller proportions. The venom is so fast-acting that the snake runs little risk of its prey fighting back and causing an injury.

It is unusual amongst snakes in changing color slightly according to the season, becoming paler in summer and darker in winter, presumably to regulate the amount of heat it absorbs.

Its alternative name of Fierce Snake is derived from an earlier description in which it was named *Diemania ferox* (*ferox* meaning fierce), chosen because of the snake's visual appearance rather than its demeanor.

→ The Inland Taipan is very rare and lives in a remote part of Australia. For almost 100 years it was thought to have been extinct, until it was rediscovered in 1972.

BOTHROPS ASPER

Terciopelo

A well-hidden hazard

SCIENTIFIC NAME	*Bothrops asper* (Garman, 1883)
FAMILY	Viperidae
SUBFAMILY	Crotalinae
SIZE	4–6 ft (1.2–1.8 m), exceptionally to 8 ft (2.5 m)
REPRODUCTION	Oviparous, with up to 80 young
HABITAT	Tropical rainforests, plantations

A common snake and greatly feared where it is found owing to its powerful venom, unpredictable temperament and tendency to rest along paths, riverbanks, and other areas frequented by humans. By day it coils among vegetation or dead leaves, where it is well camouflaged and can easily be overlooked and trodden on.

In Costa Rica, and most other countries within its range, this species is responsible for about half of all venomous snake bites, and it is the cause of most deaths, a serious problem among people who work in fields and plantations, many of whom wear flimsy footwear, if they wear any at all.

On the other hand, it is an efficient hunter of rats, which is undoubtedly why it has a tendency to gravitate toward plantations and human habitation. It is primarily an ambush predator and may return to the same position for several days in a row if it senses that prey is likely to pass within range. When mammals are not available, Terciopelos switch to feeding on amphibians, which probably explains their preference for sites near water. Juveniles feed almost exclusively on frogs, and have yellow tips to their tails that a re thought to act as lures.

Its common name Terciopelo means "velvet skin," and it is also known as the "Fer-de-Lance" (lancehead) in some regions—although this name is more accurately applied to the closely related *Bothrops atrox*, a species with which the Terciopelo has frequently been confused in the past. The name Fer-de-Lance may also be applied to members of the genus *Bothrops* as a whole.

→ The Terciopelo is among the most dangerous snakes in Central America because it is very common around plantations and villages, well camouflaged, and quick to strike.

Brown Tree Snake

A destructive interloper

SCIENTIFIC NAME	*Boiga irregularis* (Bechstein, 1802)
FAMILY	Colubridae
SUBFAMILY	Colubrinae
SIZE	Up to 6½ ft (2 m)
REPRODUCTION	Oviparous
HABITAT	Varied, from primary and secondary rainforest and mangrove forests to plantations and urban gardens

A highly arboreal species that will occasionally descend to ground level, this is an adaptable snake that has wrought havoc with the indigenous bird populations on Guam, a remote island in the western Pacific region.

Boiga irregularis is a mildly venomous back-fanged snake from northeastern Australia and numerous large and small islands in the South Pacific. In the late 1940s it was accidentally introduced to the island of Guam, thought to have arrived as a stowaway with goods supplied to the United States military base there. It spread rapidly, and by 1982 it had colonized all the suitable forest habitat on the island.

In the 1960s it was noticed that several native bird populations on Guam were declining, and by the late 1980s several species, including the Guam Flycatcher, *Myiagra freycineti*, and the Guam Rail, *Rallus owstoni*, were declared extinct on the island, while others were reduced to small remnant populations. Experiments using captive

Quail, *Coturnix coturnix*, proved that the snakes were responsible for these declines.

Lizards, mostly skinks and geckos, were also being eaten, and the snakes took domestic chickens, entered houses, and even attempted to eat sleeping infants. In addition, the snakes were causing frequent power cuts by climbing onto overhead cables. The lack of natural predators has been blamed for the exceptional impact the snake has had on native wildlife, together with populations of native birds that had no defense against the snake, and a forest structure that happened to favor the snake to the detriment of the birds.

In 2021, $4.1 billion was earmarked for control of the Brown Tree Snake on Guam. The government of Hawaii, which has no native snakes, and which is fearful of a similar invasion leading to the elimination of several rare endemic birds, has introduced fines of up to $200,000 and three years in prison for owning a snake of any kind on the islands.

→ Implicated in the extinction of at least two bird species on Guam, where it was introduced accidentally, the Brown Tree Snake is native to Australia.

Antiguan Racer

Back from the brink

SCIENTIFIC NAME	*Alsophis antiguae* Parker, 1933
FAMILY	Colubridae
SUBFAMILY	Dipsadinae
SIZE	Up to 3¼ ft (1 m)
REPRODUCTION	Oviparous
HABITAT	Rocky outcrops and scrub

The Antiguan Racer is a major conservation success story, with populations growing from a nucleus of an estimated 50 surviving individuals in 1995 to around 1,000 in 2015. The original population on Antigua was exterminated firstly by Black Rats, *Rattus rattus*, which were introduced accidentally, followed by mongooses, which were introduced deliberately in an ill-fated attempt to control the rats but which also preyed on the racers.

In addition, land clearance for agriculture had already had a negative impact on the snakes, and they were thought to be extinct on the main island of Antigua. They were rediscovered, however, on a neighboring small island, Great Bird Island, and efforts were made to rid the island of its rat population by poisoning. Once this was achieved, the snake population grew rapidly, until the small island reached capacity as the numbers of the lizards on which the snake fed was depleted. At this point, snakes were introduced to additional predator-free offshore islands, where more snake populations quickly became established.

At the same time, efforts were made to persuade local human communities to value the snakes as a special and unique feature of their islands. Films and television documentaries were made to raise the profile of the project, equipment for tracking the snakes was purchased, and local people took on the role of monitoring them. This is a fine example of how snake populations can recover, given some basic fieldwork and the will to make a difference.

→ The future of the Antiguan Racer is looking a lot brighter since measures were taken to eliminate predators and move the snakes to a safer island in the late 1990s.

Burmese Python

An unwelcome guest

SCIENTIFIC NAME	*Python bivittatus* Kuhl, 1820
FAMILY	Pythonidae
SIZE	Up to 16 ft (5 m), occasionally longer
REPRODUCTION	Oviparous
HABITAT	Grassland, marshes, swamps, and forest clearings

One of the "giant" snakes, though not the biggest. The Burmese Python is a familiar sight in zoos, as well as in the hands of pet keepers—not to mention exotic dancers. It is impressive yet easily tamed and in all its various color forms.

The markings of this snake are beautiful, and always include a dark, arrow-shaped mark on the nape of the neck. This is a powerful constrictor and will eat prey up to the size of domestic pigs and goats, as well as wild mammals and birds. It has heat pits in a small number of its upper labial scales, directed forward and enabling it to detect prey in darkness, making it an efficient hunter. It also enters water regularly and is an excellent swimmer. A most adaptable snake.

Unfortunately, from an environmental point of view, it is too adaptable. Burmese Pythons were introduced to Florida in the late 1970s, presumably as unwanted household pets when they grew too large. They quickly took to Florida, with its warm, humid climate, network of rivers and canals, and plentiful amounts and varity of wildlife. The growing population of pythons was centered on the Everglades National Park; by 2020 there were estimated to be 150,000 there, and their range extended to 1,000 sq miles (2,500 sq km).

In 2022 a 17 ft (5.2 m) python was caught in Florida and weighed 216 lb (98 kg), and in 2023 this was exceeded by one measuring 19 ft (5.8 m). The 17 ft individual was a female containing 122 developing eggs. A clutch of 111 incubating eggs has also been found. Prey taken from the stomachs of Burmese Pythons in Florida include 24 species of mammals, 47 species of birds, and two species of reptiles (see also page 243).

A program aimed at eliminating the species from Florida, by trapping and manually catching them, has met with only limited success even though 16,000 are thought to have been removed since 2000. The chances of eliminating the snakes completely are minimal.

→ A familiar pet, the Burmese Python has become infamous in recent times as a highly successful, but unwanted, introduced species in Florida, and possibly elsewhere.

Aesculapian Snake

The snake that heals itself

SCIENTIFIC NAME	*Zamenis longissimus* (Laurenti, 1768)
FAMILY	Colubridae
SUBFAMILY	Colubrinae
SIZE	4–5 ft (1.2–1.5 m), occasionally longer
REPRODUCTION	Oviparous
HABITAT	Woods, wooded hillsides, scrub, field edges

An elegant and graceful snake that climbs well and has a long association with humans. Aesculapian Snakes are named for the Greek god Asklepios (known to the Romans as Aesculapius), son of Apollo and the god of medicine and healing.

The medical symbol, consisting of a staff with a single serpent entwined, is known as the rod of Asclepius. It should not be confused with the *caduceus*, the symbol of Hermes, in which the staff has two snakes coiled around it.

The reason for the belief in the curative powers of the snake is thought to lie in its regular shedding of the outer layer of its skin (the epidermis) to expose the new, shiny, undamaged skin beneath. This is likened to a "rebirth" in several religions, hence the association with healing.

The Aesculapian Snake has a patchy distribution in central Europe, with several outlying populations in Germany and the northern Czech Republic in addition to its main range. In the past it has been proposed that these may have originated from snakes released into temples in those areas, but it is more likely that they are relict populations left behind when the region was settled and developed. In addition to its natural range, small populations occur in north Wales and central London in the United Kingdom, the descendants of escapees or deliberate introductions from nearby zoos.

Females lay up to 15 elongated eggs in rotting vegetation, tree stumps, and similar places, and they hatch after about two months. Juvenile Aesculapian Snakes are very different from the adults, having dark stripes on the head, yellow "collars," and a series of dark blotches down their backs.

→ The adaptable Aesculapian Snake has a patchy distribution in Europe but can be common in suitable habitats.

PANTHEROPHIS GUTTATUS

Corn Snake

A favorite pet snake

SCIENTIFIC NAME	*Pantherophis guttatus* (Linnaeus, 1766)
FAMILY	Colubridae
SUBFAMILY	Colubrinae
SIZE	2½–5 ft (80–160 cm), occasionally to 6 ft (180 cm)
REPRODUCTION	Oviparous
HABITAT	Woods, grasslands and scrub, and urban environments

The Corn Snake, also known as the Red Ratsnake, is arguably the most popular "pet" snake. It is bred in large numbers, and various color forms, to satisfy the demand from amateur snake-keepers. Many budding herpetologists will have started their journey with a pet Corn Snake.

Corn Snakes tend to be nocturnal, resting by day in burrows, or under objects on the ground, including boards and trash, but they also climb well, into trees and abandoned buildings. They feed mostly on warm-blooded prey when adult, although juveniles also eat lizards and frogs. Hatchling size varies, and some are probably too small to take even newborn rodents.

There is a natural regional variation in color and markings, with those coming from the south of Florida tending toward orange-red saddles on a silver-gray background, whereas those from further north have rich red saddles, outlined in black, on an orange background.

"Okeetee" Corn Snakes from South Carolina are often considered the most colorful.

Because Corn Snakes are relatively easy to breed in captivity, and can be very prolific, selective breeding has produced a vast array of color forms or "morphs." Many are based on missing colors, such as the "amelanistic" forms, which lack all black markings but still retain the red pigmentation, and "anerythristic" snakes, in which the red pigment is missing. In other forms the saddles down the back are replaced by a single or double stripe. By combining and back-crossing the various forms, a myriad of colors and patterns can be created. Animals so produced are sometimes given fanciful names, often to help market them, one suspects.

Due to their popularity, coupled with their ability to escape from their cages, Corn Snakes occasionally turn up where they're not wanted. Mostly these are singletons and no lasting harm is done, but there is evidence that they are becoming established in parts of Australia, notably in the Sydney area.

→ The Corn Snake or Red Ratsnake is one of the most beautiful North American species, and familiar in the pet trade, in all its various guises.

GLOSSARY

Adaptation A behavioral or physical characteristic that has evolved to improve an organism's chances of survival in a particular environment.

Ambush predator A predator that waits in a concealed position until prey comes within range. Many are well camouflaged and they may enhance their chances by using a lure. Also known as a "sit-and-wait" predator.

Aposematic coloration The use of bright colors to warn predators that an animal is dangerous, toxic, or distasteful (or pretending to be).

Arboreal Living in trees.

Autotomy *See caudal autotomy.*

Back-fanged snake *See rear-fanged snake.*

Bask To move the body to a place where it can acquire heat, usually from the sun, although some snakes bask by positioning themselves under flat rocks, etc.

Binocular vision The ability to focus both eyes on an object so that its distance can be judged.

Brille Also known as the spectacle, this is the transparent scale covering the eyes of snakes.

Cannibalism Eating members of one's own species (but not eating a different species in the same order, so snakes that eat other types of snakes are not cannibalistic, for instance).

Caudal Relating to the tail.

Caudal autotomy Discarding the tail in self-defense. Only a few snakes are known to practice this.

Chromatophore Cell containing pigment within the skin. Chromatophores give animals their colors and patterns.

Class A taxonomic unit containing one or more *orders*, but below the hierarchical unit of phylum.

Cloaca Common opening of the excretory and reproductive systems.

Columella A small bone that transmits vibrations to the inner ear (equivalent to the stapes in mammals).

Constrictor A snake that kills its prey by coiling around it tightly until it suffocates.

Convergent evolution The process by which species come to resemble each other because they have adapted to similar environmental conditions.

Crepuscular Active at dusk and/or dawn.

Cryptic coloration Color or pattern that breaks up the outline of an animal, helping it to blend in with its surroundings and making it difficult to see.

Cryptic species Two or more species that look identical but which differ genetically and may be genetically isolated through courtship behavior.

Desert A region that experiences little rainfall. Deserts can consist of completely barren sand dunes or fairly heavily vegetated regions in which the plants are drought-adapted. Deserts are often home to many snakes.

Dimorphism Existing in two different forms. For instance, if males and females are different in shape, size, or color they are said to be sexually dimorphic. *See also polymorphism.*

Diurnal Active by day.

Dorsal Referring to the back. Opposite of *ventral*.

Ectotherm An animal, such as a snake, that relies on outside sources to maintain a suitable body temperature. Often (inaccurately) referred to as "cold-blooded."

Epidermis Surface layer of skin; in snakes, it is the epidermis that is discarded when the snake sheds its skin.

Extinct There are no known individuals remaining in the wild or in captivity. Several snakes have become extinct in recent years, and more are likely to disappear in the next couple of decades. A species is said to be "locally extinct" when a particular population has become extinct even though it may survive elsewhere.

Extinct in the Wild An *IUCN* category indicating a species that exists only in captivity.

Family A taxonomic unit containing one or more genera, but below *order* in the taxonomic hierarchy. In zoology, family names end in –idae.

Fang Long tooth, often grooved or hollow in the case of venomous snakes.

Fossorial Relating to a burrowing lifestyle.

Genus (pl. genera) A taxonomic unit that contains one or more closely related species. Genera come between *species* and *families* in *taxonomic* hierarchy. Species belonging to the same genus have the same first name – the generic name – which must be unique. The generic name begins with a capital letter and is written in italics. Examples of genera are *Python* and *Crotalus*.

Heat-sensitive pits (or heat pits) Organs found in the faces of some snakes, used for detecting small temperature changes, such as the radiant heat given off by warm-blooded animals. In boas and pythons they are situated within or between the *labial* (lip) scales, and in pit vipers they are paired, between the eyes and the nostrils.

Hemipenis (pl. hemipenes) One of the paired reproductive organs of male snakes.

Hibernation An extended period of torpor during which an animal's metabolism slows down. Snakes from cool regions often hibernate for several months in the winter. Sometimes known as brumation.

Internal fertilization Method of reproduction in which the fusion of eggs and sperm takes place inside the female's body. All snakes have internal fertilization.

Introduced species A species that has been introduced to a part of the world where it does not naturally occur, either deliberately, such as in the case of the Cane Toad, which was introduced to Australia, or by accident, as in the introduction of the Brown Tree Snake to the island of Guam.

Iridescent Colors that appear to change according to the direction of the light. Some snakes have particularly iridescent scales.

Iridophores Cells in the skin that reflect light, often producing blue coloration by a process known as Tyndall scattering. If the layer of iridophores is overlaid by yellow cells (*xanthophores*) these will turn the blue coloration to green.

IUCN International Union for Conservation of Nature, the organization that assesses the threats to wildlife on a species-by-species basis.

IUCN Red List A list prepared by the IUCN that identifies the conservation status of each species. Species range from Least Concern, Near Threatened, Vulnerable, Endangered, Critically Endangered, to *Extinct*. Many snakes are Data Deficient (DD) because they have not yet been assessed.

Jacobson's organ In snakes, a pair of organs that open onto the roof of the mouth and connect to the olfactory part of the brain. The forked tongue of snakes picks up scent particles from the environment and passes them to the Jacobson's organ, which then transfers the information to the brain.

Keeled scales Scales that have one or more longitudinal ridges, usually down the center.

Labial scales Scales bordering the mouth.

Lateral Referring to the side. Laterally compressed, for instance, means flattened from side to side (i.e., higher than wide), as in the cross-section of some arboreal snakes and the tails of sea snakes.

Lingual fossa Notch in the upper jaw through which the snake extends its tongue.

Lure Lures are used by some *ambush predators* to entice their prey to move within range, such as a brightly colored tail that the snake moves to simulate a grub or caterpillar.

Maxilla One of the bones forming the upper jaw.

New World North and South America and their associated islands. *See also Old World.*

Nomenclature A system of naming plants and animals, in which certain agreed rules must be followed.

Nominate subspecies If there is more than one subspecies, the nominate subspecies is the one that represents a species when it was first described. The subspecific name of the nominate subspecies repeats the specific name, so *Vipera aspis aspis* is the nominate subspecies of *Vipera aspis* (the Asp Viper). Other subspecies described later must have different subspecific names, such as *Vipera aspis montecristi*.

Old World Europe, Asia, Africa, and Australasia, and their associated islands. *See also New World.*

Olfactory Referring to the sense of smell. A form of chemical communication.

Order A taxonomic unit containing one or more *families* but below the level of *class*.

Oviparous Egg-laying.

Ovo-viviparous Reproducing by means of eggs that are retained inside the mother until they hatch, but which do not receive any nourishment from the mother. Most live-bearing snakes are ovo-viviparous. *See also viviparous.*

Palatine One of the bones forming the roof of the mouth.

Parthenogenesis Referring to a species in which females can produce viable offspring without the need to mate with a male. Only one species of snake is fully parthenogenic (*see* page 134).

Pectoral girdle The bones supporting the front limbs (or fins), such as the shoulder bones, absent in all snakes.

Pelvic girdle The bones supporting the hind limbs in vertebrates. Some snakes have retained vestigial pelvic girdles.

Pheromone A substance secreted or excreted by an organism that produces a response in another individual of the same species. A form of chemical communication.

Pigment A substance that gives an animal its color. In snakes, pigments are embedded in cells in the skin or scales.

Pleurodont Describing teeth that are fused only to the inner surface of the jawbones, the arrangement found in all snakes. *See also thecodont.*

GLOSSARY

Polymorphism Existing in two or more different forms within a population. Species may be polymorphic to avoid predation, for instance, because predators tend to use a search image to find prey, and by being different in color or pattern, a proportion of individuals may escape notice. *See also dimorphism.*

Precaudal The portion of the body in front of the tail.

Prehensile (tail) A tail that can be used to grip, as in many *arboreal* snakes.

Pterygoid Paired bones forming part of the palate.

Quadrate bone A small bone at the back of the skull.

Rainforest Forest, usually in the tropics, which has high rainfall and high humidity. Tropical rainforests are especially rich in species of snakes.

Rattle Structure on the tail of *rattlesnakes,* consisting of loosely interlocking segments of dead skin, which makes a sound when vibrated.

Rattlesnakes Vipers of the *genera Crotalus* and *Sistrurus* that have a warning rattle on the tip of their tail.

Rear-fanged snake Species in which enlarged *fangs* are positioned toward the rear of the mouth. They may be used to introduce *venom,* and some, such as the Boomslang, are dangerous to humans.

Rostral scale The scale at the tip of a snake's snout, sometimes modified for digging.

Scalation The size, shape, and arrangement of scales.

Species A population of potentially interbreeding individuals that are reproductively isolated from other such groups. The species is the basic unit of *taxonomic* classification. Specific names always begin with a lowercase letter, even if they are based on a name or place (e.g., *darwini*), and are written in italics.

Spectacle *See brille.*

Spur Pointed structures in pythons and boas, positioned each side of the cloaca and forming *vestigial* limbs, sometimes used in courtship.

Subadult Stage of life between a juvenile and an adult.

Subspecies A *taxonomic* unit that is subordinate to *species.* Different subspecies within a species are capable of breeding with each other but do not do so under natural conditions because subspecies are geographically separated. Not all biologists recognize the subspecies concept.

Supraocular scales The scales above the eyes. Their size and shape are sometimes useful in species identification.

Taxonomy The arrangement of plants and animals into groups based on their natural (evolutionary) relationships.

Terrestrial Living on the ground.

Thecodont Describing teeth that are embedded into sockets in the jawbones. Snakes do not have thecodont teeth, but some other reptiles do. *See also pleurodont.*

Thermoregulation The regulation of the body temperature by physiological or behavioral means to ensure that it remains within a preferred temperature range whenever possible.

Tubercle A small fleshy pimple or protuberance.

Tuberculate Covered in tubercles.

Venom Fluid containing toxins that are injected into prey by snakes, for example.

Ventral Referring to the underside, as in "ventral scales." Opposite of *dorsal.*

Vestigial Describing part of an animal that is in the process of being lost through the evolutionary process and is small and often serves no function. *See also spur.*

Viviparous Giving birth to live young which develop in, and are nourished by, the mother. Most live-bearing snakes are actually *ovo-viviparous,* but a small number are viviparous.

Xanthophore A type of *chromatophore* that contains yellow *pigment,* usually found in the uppermost layer of the skin or dermis.

RESOURCES

BOOKS

Campbell, J. A. and Lamar, W. W. *The Venomous Reptiles of Latin America.* Cornell University Press, Ithaca and London, 1989.

Egan, D. *Snakes of Arabia.* Motivate Publishing, Dubai, 2007.

Ernst, C. H. and Ernst, E. M. *Snakes of the United States and Canada.* Smithsonian Books, Washington DC, 2003.

Geniez, P. *Snakes of Europe, North Africa and the Middle East.* Princeton University Press, Princeton, 2018.

Glaw, F. and Vences, M. *A Fieldguide to the Reptiles and Amphibians of Madagascar,* 3rd edition. Published by the authors, Germany, 2007.

Greene, H. W. *Snakes: the Evolution of Mystery in Nature.* University of California Press, Berkeley, 1997.

Grismer, L. L. *Amphibians and Reptiles of Baja California.* University of California Press, Berkeley, 2002.

Heatwole, H. *Sea Snakes.* New South Wales University Press, Sydney, 1987.

Kreiner, G. *The Snakes of Europe.* Edition Chimaira, Frankfurt, 2007.

Lillywhite, H. B. *How Snakes Work.* Oxford University Press, Oxford, 2014.

Mattison, C. *Snake: the Essential Visual Guide,* revised edition. Dorling Kindersley, London, 2015.

Milton, N. *The Secret Life of the Adder.* White Owl (Pen and Sword Books), Yorkshire and Philadelphia, 2022.

O'Shea, M. *A Guide to the Snakes of Papua New Guinea.* Independent Publishing, Port Moresby, 1996.

Phelps, A. *Old World Vipers.* Edition Chimaira, Frankfurt, 2010.

Savage, J. M. *The Amphibians and Reptiles of Costa Rica.* University of Chicago Press, Chicago, 2006.

Spawls, S. and Branch, W. R. *The Dangerous Snakes of Africa.* Princeton University Press, Princeton, 2020.

Steubing, R. B. and Inger, R. F. *A Field Guide to the Snakes of Borneo.* 2nd edition. Natural History Publications (Borneo), Kota Kinabalu, Malaysia, 2014.

Whitaker, R. and Captain, A. *Snakes of India.* Draco Books, Chennai, 2004.

Wilson, S. and Swan, G. *A Complete Guide to the Reptiles of Australia,* 6th edition. New Holland, Sydney, 2020.

ONLINE ARTICLES AND OTHER RESOURCES

The Internet is a huge fund of information about snakes, and it is impossible to list even a small proportion of what is available, especially as many articles come and go. The following are a few of those that I find especially interesting and useful, and which may not crop up in general searches for snakes.

Arteaga, A., Pyron, R. A., Batista, A., *et al.* Systematic revision of the Eyelash Palm-Pitviper *Bothriechis schlegelii* (Serpentes, Viperidae), with the description of five new species and revalidation of three. *Evolutionary Systematics* 2024, 8(1): 15–64. https://doi.org/10.3897/evolsyst.8.114527

Bohm, M. *Reptiles and climate change.* Zoological Society of London, 2016. www.zsl.org/news-and-events/feature/reptiles-climate-change

Elfes, C. T., Livingstone, S. R., Lane, A., *et al.* Fascinating and forgotten: the conservation status of marine elapid snakes. *Herpetological Conservation and Biology* 2013, 8(1): 37–52. www.herpconbio.org/contents_vol8_issue1.html

Fathinia, B., Rastegar-Pouyani, N., Rastegar-Pouyani, E., Todehdehghan, F., and Amiri, F. Avian deception using an elaborate caudal lure in *Pseudocerastes urarachnoides* (Serpentes: Viperidae). *Amphibia-Reptilia* 2015, 36(3), 223–231. https://doi.org/10.1163/15685381-00002997

Fiorvanti, C. 2018. How snakes arrived in the Galapagos. *Pesquisa,* issue 270. https://revistapesquisa.fapesp.br/en/how-snakes-arrived-in-the-galapagos

Hagman, M., Phillips, B. L., and Shine, R. Tails of enticement: caudal luring by an ambush-foraging snake (*Acanthophis praelongus*, Elapidae). *Functional Ecology* 2008, 22 (6): 1134–1139. https://doi.org/10.1111/j.1365-2435.2008.01466.x

International Union for Conservation of Nature. *The IUCN Red List of Threatened Species.* A searchable database with notes on the listed species. Updated regularly. www.iucnredlist.org

King Cobra Conservancy. *Living with the King.* An excellent short film describing conservation of the King Cobra in India. www.thekingcobra.org/living-with-the-king

Loss, S. R., Will, T., and Marra, P. P. The impact of free-ranging domestic cats on wildlife of the United States. *Nature Communications* 2013, 4, 1396. https://doi.org/10.1038/ncomms2380

Save the Snakes. *Global snake conservation.* Links to several conservation organizations around the world. https://savethesnakes.org/conservation

Uetz, P., Freed, P., Aguilar, R., Reyes, F., Kudera, J. & Hošek, J. (eds.) *The Reptile Database.* An invaluable resource, updated regularly. www.reptile-database.org.

INDEX

I would like to thank Kate Shanahan, senior editor at UniPress, and Robert Kirk, publisher at Princeton University Press, for giving me the opportunity to write this book. The project manager, David Price-Goodfellow, was professional, knowledgeable, and patient throughout, and his guidance has been indispensable. The managing editor, Slav Todorov, has overseen all aspects of the project and has stepped in to help on several occasions.

Thanks are also due, in alphabetical order, to Wayne Blades (designer), Robert Brandt (illustrator), Hugh Brazier (copy editor), Tom Broadbent (picture researcher), Jason Hook (publisher), and to an anonymous reviewer. The book would not have been possible without the contributions of these people.

It is a pleasure to thank Cristian Torica, Alan Francis, and Lewis Burnett for permission to use some of their superb photographs, which have added greatly to the book's visual appeal, and to acknowledge the contributions of many agency photographers. Dave Howard and his enthusiastic team at Rainforest Exotics kindly allowed me to photograph several of their beautiful snakes.

And finally, I am grateful to my wife Gretchen, and the friends and colleagues, too numerous to mention, who have accompanied me on photographic and research trips to many parts of the world over many years. Your help, companionship, and good nature have always been appreciated.

PICTURE CREDITS

Guide: TL Top left, ML Middle left, BL Bottom left, TR Top right, MR Middle Right, BR Bottom right.
2 Sergey and Marina Pyataev/Shutterstock, 3 Dr Morley Read/Shutterstock, 4–5 Clockwise L–R Benny Marty, Nynke van Holten x 2, asawinimages, Eric Isselee, Aastels, Agus_Gatam, Kurit afshen (All Shutterstock), 6–8 Chris Mattison, 9 Top Christophe Courteau/Nature Picture Library/Alamy Stock Photo, 9 Bottom WaterFrame/Alamy Stock Photo, 10–11 model by Kevin Hockley, artist/photo by James Di Loreto, Smithsonian Institution, 12 Top Patrick K. Campbell/Shutterstock, 12 Bottom & 14 Chris Mattison, 15 Top Cristian Torica, 15 Bottom Chris Mattison, 16 (both) Chris Mattison, 17 TL Brad Leue/Alamy Stock Photo, 17 TR and Bottom Chris Mattison, 18 Chris Mattison, 19 Top Cristian Torica, 19 Bottom Chris Mattison, 20 Cristian Torica, 21–23 Chris Mattison, 27 Oliver Tookey/Alamy Stock Photo, 28–29 Chris Mattison, 30 Cristian Torica, 32 Chris Mattison, 33 Craig Cordier/Shutterstock, 35 Bottom Cristian Torica, 35 Both insets Chris Mattison, 36 and 37 Left Cristian Torica, 37 Right Chris Mattison, 38–39 Chris Mattison, 40 Cristian Torica, 41 Top Dr Morley Read/Shutterstock, 41 Middle PetlinDmitry/Shutterstock, 41 Bottom Eric Isselee/Shutterstock, 42 Lauren Suryanata/Shutterstock, 43 Cristian Torica, 45 Chris Mattison, 45 Inset Cristian Torica, 46 Both Chris Mattison, 47 Both Cristian Torica, 48 Chris Mattison, 49 Top Chris Mattison, 49 Bottom Christian Torica, 50 Inset Chris Mattison, 51 Clement Carbillet/Biosphoto/Alamy, 52 & 53 Top Chris Mattison, 53 Bottom GoodFocused/Shutterstock, 54 Daniel Heuclin/Naturepl.com, 56–58 Left Chris Mattison, 58 Right Cede Prudente/Avalon.red/Alamy Stock Photo, 59 Yvette Cardozo/Alamy Stock Photo, 60 Lee Yiu Tung/Shutterstock, 61 Chris Mattison, 63 Fauzan Maududdin/Shutterstock, 65 blickwinkel/B. Trapp/Alamy, 67–77 Chris Mattison, 79 Rusty Dodson/Shutterstock, 81 Tim Laman/Naturepl.com, 84–87 Chris Mattison, 88 Jay Ondreicka/Shutterstock, 89 Left JeanMaurice/Shutterstock, 89 Right Lou Staunton, 90 Top Robert Hamilton/Alamy Stock Photo, 90 Bottom Dan_Koleska/Shutterstock, 91 Inset Gavin Maxwell/Naturepl.com, 91 Bottom Ton Bangkeaw/Shutterstock, 92–93 Chris Mattison, 94 Paul D Stewart/Nature PL/Alamy, 95 Left Chris Mattison, 95 Right Michael Nolan/robertharding/Alamy Stock Photo, 96 Matt from Melbourne, Australia, CC BY 2.0, 97 Lewis Burnett, 98 kristian bell/RooM the Agency/Alamy Stock Photo, 99 Colin N. Perkel/Alamy Stock Photo, 101–103 Chris Mattison, 105 Gabriel Rojo/Naturepl.com, 107 Alain Dragesco-Joffe/Naturepl.com, 109 Martin Willis/Naturepl.com, 111 Daniel Heuclin/Nature PL/Alamy, 113 Tanguy de Saint-Cyr/Shutterstock, 115 Claudio Contreras/Naturepl.com, 117 Jean-Paul Ferrero/AUSCAPE/Alamy Stock Photo, 120 Jelger Herder/Buiten-Beeld/Alamy Stock Photo, 121 Tahmid Hasan Sobuj/Shutterstock, 122 Francois Savigny/Naturepl.com, 123 Lorraine Swanson/Shutterstock, 124–125 Chris Mattison, 126 Alan Francis, 127 dpa picture alliance/Alamy Stock Photo, 128 Top Chris Mattison, 128 Bottom Robert Gill/apilio/Alamy Stock Photo, 129 Chris Mattison, 130–131 Chris Mattison, 132 E.R. Degginger/Alamy Stock Photo, 133 Inset Tony Phelps/Naturepl.com, 133 Pedro Luna/Shutterstock, 134 Chris Mattison, 135 Candice Davis/Missouri Department of Conservation via AP/Alamy, 136 Left Daniel Heuclin/Biosphoto/Alamy Stock Photo, 136 Right Joe Blossom/Alamy Stock Photo, 137 Mark_Kostich/Shutterstock, 138 Top Donald M. Jones/Shutterstock, 138 Bottom Chris Mattison, 139 Lauren Suryanata/Shutterstock, 141 GFC Collection/Alamy Stock Photo, 143 Chris Mattison, 145 Nate Chappell/BIA/Minden Pictures/Alamy, 147 Ray Wilson/Alamy Stock Photo, 149 Jan Hejda/Shutterstock, 151 Thomas Marent/Minden Pictures/Alamy, 154 Chien Lee/Naturepl.com, 155 Gulliver20/Shutterstock, 156–157 Gunter Nuyts/Shutterstock, 158 Helmut Göthel Symbiosis/Alamy Stock Photo, 158 Top Chien Lee/Naturepl.com, 158 Bottom Dmitry Eagle Orlov/Shutterstock, 160 & 161 Top Chris Mattison, 161 MR Cristian Torica, 161 BR Rusty Dodson/Shutterstock, 162 Que Images/Shutterstock, 163 Cristian Torica, 164 Anthony Bannister/Avalon.red/Alamy Stock Photo, 165 Jelger Herder/Buiten-Beeld/Alamy Stock Photo, 166 Tui De Roy/Nature Picture Library/Alamy Stock Photo, 167 Mufti Adi Utomo/Shutterstock, 168 Martha Holmes/Naturepl.com, 169 tatui suwat/Shutterstock, 170 (both) Chris Mattison, 171 Alberto Loyo/Shutterstock, 172–174 Chris Mattison, 175 Top reptiles4all/Shutterstock, 175 Bottom Ken Griffiths/Shutterstock, 176 Chris Mattison, 177 Sheril Kannoth/Shutterstock, 179 Cristian Torica, 181–183 Chris Mattison, 185 Cristian Torica, 187 Claudio Contreras/Naturepl.com, 189 Blue Planet Archive/D. R. Schrichte, 191–195 Chris Mattison, 197 Alex Mustard/Naturepl.com 200 Paulpixs/Shutterstock, 201 Top Lou Coetzer/Naturepl.com, 201 Bottom & 202–204 Chris Mattison, 205 David A Litman/Shutterstock, 206 Cristian Torica, 207 Left Chris Mattison, 207 Right Cristian Torica, 208 Left Michael & Patricia Fogden/Minden Pictures/Alamy, 208 Right RealityImages/Shutterstock, 209 Chris Mattison, 210 Andrew DuBois/Alamy Stock Photo, 211 Rusty Dodson/Shutterstock, 212 Top El Golli Mohamed, CC BY-SA 4.0 https://creativecommons.org/licenses/by-sa/4.0, via Wikimedia Commons, 212 Inset Chris Mattison, 213 Top Chase D'animulls/Shutterstock, 213 Bottom Federico.Crovetto/Shutterstock, 214–215 Stu Porter/Shutterstock, 217 Roger de la Harpe/Alamy Stock Photo, 219 Scott Trageser, 221 Chris Mattison, 223 Cristian Torica, 225 Chris Mattison, 227 Guy Edwardes/Nature Picture Library/Alamy Stock Photo, 229 Chris Mattison, 231 Chris Mattison, 233 Chris Mattison, 236 Chris Mattison, 237 Top Rogers Fund, 2007 Met Museum, 237 BL Rogers Fund, 1918 Met Museum, 237 BR Theodore M. Davis Collection, Bequest of Theodore M. Davis, 1915 Met Museum, 238–239 Abbie Warnock-Matthews/Shutterstock, 240 Elbridge Ayer Burbank/Library of Congress, 241 Top Ultrasto/Shutterstock, 241 Bottom LEROY Francis/hemis.fr/Alamy, 242 Narelle Power/Alamy Stock Photo, 243 Aaron Gekoski/World Animal Protection/Naturepl.com, 244–248 Left Chris Mattison, 248 Right Riadi Pracipta66/Shutterstock, 249 PARALAXIS/Shutterstock, 250 Jorge Fernandez/Alamy Stock Photo, 251 Left Deco/Alamy Stock Photo, 251 Right OlgaOvcharenko/Shutterstock, 252 Chris Mattison, 253 Top Eva Pruchova/Alamy Stock Photo, 253 Middle K-FK/Shutterstock, 253 Bottom Ethan Daniels/Shutterstock, 254 taviphoto/Shutterstock, 255 Public Domain, 256 Lewis Burnett, 257 Blue Planet Archive/Ethan Daniels, 258 Blue Planet Archive/Michael Weberberger, 259 Joseph Cobaleda/Shutterstock, 260 Kamal Devkota, Save The Lives (www.savethesnakes.org/nepal), 261 Nik Cole/Durrells.org, 263 Thomas Marent/Naturepl.com, 265 Thomas Marent/Naturepl.com, 267 Chris Mattison, 269 Bruce Thomson/Naturepl.com, 271 John Cancalosi/Alamy Stock Photo, 273 Edwin Giesbers/Naturepl.com, 275 Chris Mattison, 277 John Cancalosi/Alamy Stock Photo.